Service Integration and Management (SIAM™) Professional Body of Knowledge (BoK)

Second edition

Service Integration and Management (SIAM™) Professional Body of Knowledge (BoK)

Second edition

MICHELLE MAJOR-GOLDSMITH, SIMON DORST,
CLAIRE AGUTTER ET AL.

IT Governance Publishing Ltd
Unit 3, Clive Court
Bartholomew's Walk
Cambridgeshire Business Park
Ely, Cambridgeshire
CB7 4EA
United Kingdom
www.itgovernancepublishing.co.uk

First edition published in 2018 by Van Haren Publishing.

Second edition published in the United Kingdom in 2021 by IT Governance Publishing

ISBN 978-1-78778-313-3

ACKNOWLEDGEMENTS

About this book

Scopism would like to thank the following people and organizations for their contributions to this book:

Lead architects

Michelle Major-Goldsmith, Kinetic IT

Simon Dorst, Kinetic IT

Contributing authors and reviewers

Alison Cartlidge, Sopra Steria

Andrea Kis, Independent

Angelo Leisinger, CLAVIS klw AG

Anna Leyland, Sopra Steria

Barry Corless, Global Knowledge

Biju Pillai, Capgemini

Caspar Miller, Westergaard

Charlotte Parnham, Atos Consulting

Christopher Bullivant, Atos

Chris Taylor-Cutter, Independent

Claire Agutter, Scopism

Damian Bowen, ITSM Value

Daniel Breston, Independent

Dave Heaton, BAE Systems

Dean Hughes, Independent

Franci Meyer, Fox ITSM South Africa

Graham Coombes, Holmwood GRP

Hans van den Bent, CLOUD-linguistics

Helen Morris, Helix SMS Ltd

Ian Clark, Fox ITSM South Africa

Ian Groves, Fujitsu

Jacob Andersen, Independent

Jan Halvorsrød, KPMG

Acknowledgements

Karen Brusch, Nationwide Building Society

Kevin Holland, Independent

Lise Dall Eriksen, BlueHat P/S

Liz Gallacher, Helix SMS Ltd

Mark Thompson, Kinetic IT

Markus Müller, ABB Information Systems

Martin Neville, Tata Consultancy Services

Matthew Burrows, BSMimpact

Neil Battell, Micro Focus

Peter McKenzie, Sintegral

Rajiv Dua, Bravemouth Consulting Limited

Sachin Bhatnagar, Kinetic IT

Sami Laurinantti, Sofigate

Samuel Santhoshkumar, Independent

Simon Hodgson, Sopra Steria

Simon Roller, BSMimpact

Stephen Howells, Kinetic IT

Steve Morgan, Syniad IT

Susan North, Sopra Steria

Tony Gray, PGDS (Prudential)

Trisha Booth, Atos

Tristan Quick, Kinetic IT

Troy Latter, 4PM Group

William Hooper, Independent

Second edition

The following volunteers contributed to the second edition of this document:

Alison Cartlidge, Sopra Steria

Andre Peppiatt, Capgemini

Anna Leyland, Sopra Steria

Biju Pillai, Capgemini

Claire Agutter, Scopism

Daniel Breston, Virtual Clarity

Ian Groves, Syamic

Acknowledgements

Julian White, Capgemini

Kevin Holland, Independent

Markus Müller, Blueponte

Martin Neville, Tata Consultancy Services

Matthew Burrows, SkillsTx/BSMimpact

Michelle Major-Goldsmith, Kinetic IT

Morten Bukh Dreier, Valcon

Pat Williams, Syamic

Reni Friis, Valcon

Richard Amster, Working-Globally

Sachin Bhatnagar, South32

Samuel Santhoshkumar, Heracles Solutions

Simon Dorst, Kinetic IT

Steve Morgan, Syniad IT

William Hooper, Oareborough Consulting

The purpose of *Service Integration and Management (SIAM™) Professional Body of Knowledge (BoK), Second edition*

The SIAM Professional Body of Knowledge (BoK) expands the description of service integration and management (SIAM) from the previously released SIAM Foundation BoK. The Foundation BoK is recommended reading before using this publication.[1]

The contents of the SIAM Professional BoK are the source material for the EXIN SIAM Professional certification.

---

[1] For more information, visit: *www.itgovernancepublishing.co.uk/product/service-integration-and-management-siam-foundation-body-of-knowledge-bok-second-edition*.

FOREWORD

To compete in the modern world of disruption and disruptors, organizations are focusing on their customers and their experiences, all of which are delivered using technology. Every company today must be – or plan the transition to – a software company, or it risks becoming digital dust left in the path of those that have! The quest for speed, quality and differentiation, which leverages technology, means organizations must focus on delivering innovation. In these ecosystems, they are becoming increasingly dependent on suppliers and partners. Management of the supplier ecosystem is critical to success – with little to no tolerance for downtime let alone failure.

This transition is not an overnight revelation, it has been evolving for years. Organizations globally commenced the shift from monolithic outsourcing engagements to multi-sourcing models during the 2000s, when they needed a process to integrate and manage these services and their suppliers. In response, SIAM models developed, primarily driven by public sector bodies, like the UK's Department for Work and Pensions, as well as innovative outsourcing users, such as General Motors (GM). The SIAM models were viewed primarily as the purview of the outsourcing community, delivering control in environments that were starting to seem unmanageable.

Unfortunately, practitioners have had little guidance or training available, which led to SIAM having a poor reputation. Today, some 10 years after the initial creation of SIAM, the management of outsourcers and suppliers is only becoming more complex with the adoption of cloud computing, the growth of the Internet of Things and the emergence of robotics.

Industrywide, SIAM guidance started in 2016 with the development of the SIAM Foundation Body of Knowledge. Now, the SIAM Professional Body of Knowledge delivers further comprehensive and consistent guidance, which the industry has sought since the inception of SIAM. Leveraging the experiences of many successful SIAM organizations and industry experts, the guidance incorporates principles which support organizations as they navigate the complexity of their growing list of suppliers, or work as suppliers within a SIAM model.

Going beyond theory, the guidance addresses the practicalities of how to establish the SIAM roadmap to effectively manage all supplier artifacts, including legacy contracts, commercial issues, security, cultural fit and behaviors, control and ownership, and of course, service level agreements.

For organizations who are facing the challenges of integration of multiple suppliers and outsourcers, this publication is highly recommended, and especially to organizations considering or already working within a SIAM model. Additionally, the publication comes highly recommended for practitioners who are looking to implement SIAM, and of course to anyone taking the SIAM Professional course and exam.

I am sure that you will find the SIAM Professional Body of Knowledge great guidance for your SIAM journey and I encourage you as you develop to give back to the community, just like those who have contributed to this publication.

Robert E Stroud CGEIT CRISC

Principal Analyst Forrester Research

About Robert Stroud (1963–2018)

Robert Stroud was a recognized industry thought leader, speaker, author and contributor to multiple best practices and standards. He drove thought leadership in the rapidly growing DevOps and continuous deployment domains.

Second edition

This second edition of the SIAM Professional BoK is a welcome development.

SIAM has never been more important than it is today. Organizations must be constantly on the front foot ready to respond quickly to change and remain ahead of the game.

In order to remain competitive and/or relevant, organizations need to be increasingly adaptable, agile, innovative and able to respond quickly to changing business requirements. This means that organizations will be looking to more (not less) suppliers that are market leaders in their areas of specialism, to provide products and services.

SIAM is needed to manage the complexity that is an increasing myriad of suppliers as a result.

This refresh incorporates changes that will make the management of this complexity easier. There is information about contracts, exploration of trust-based management, enhanced metrics and reporting, and more detail in regard to SIAM skills and capabilities. The refresh also provides more guidance in regard to the position of SIAM in environments such as DevOps and organizations moving to more agile ways of working.

The SIAM Professional BoK is the go-to source for information and subject-matter expertise on the effective management of multiple suppliers in a consistent and coherent manner.

Karen Ferris

Organisational Change Management Rebel With A Cause

About Karen Ferris

Karen has been awarded the itSMF Australia Lifetime Achievement Award, been named one of the top 25 thought leaders in service management by HDI in 2017 and 2018, and was named one of the 20 best ITSM thought leaders in 2017 by Sunview Software.

CONTENTS

CHAPTER 1: INTRODUCTION

1.1 Intended audience

This publication is designed for audience groups, including:

- Individuals wishing to build on their foundation-level knowledge of SIAM and achieve the SIAM Professional certification
- Customer organizations and their staff looking for guidance when managing a multi-service provider environment
- Service integrators and their staff wishing to work effectively in a SIAM ecosystem
- Internal and external service providers and their staff wishing to understand their role in a SIAM ecosystem
- Consultants in service management and other frameworks who wish to expand their knowledge in this area

1.2 The background of SIAM

SIAM developed in response to the challenges organizations face when using multiple service providers as part of their supply network, sometimes called multi-sourcing. Although multi-sourcing offers organizations the ability to choose the best service provider for each element of an end-to-end service, it may also incur significant management overhead and costs. Some organizations may not have the capabilities to manage service providers and their services.

The scope of SIAM

Although SIAM originated in the IT services arena, it is now used by an increasing number of organizations to manage business services.

This introduction provides a review of content from the SIAM Foundation BoK to aid understanding of the rest of the publication. It includes:

- SIAM Foundation BoK history
- SIAM terminology
- SIAM roadmap

Historically, organizations received IT services internally, using a simple structure of infrastructure and applications managed by an IT department. As technology use has become more complex, and business users have become more demanding, some organizations have chosen to outsource work between multiple service providers. This enables segregation of service elements, unlocks flexibility and reduces the risk of dependency on one service provider. Multi-sourcing also supports a 'best of breed' approach, where the organization can select services from specialized service providers.

Commissioning organizations must consider how and from where services are provided, to maximize performance of their value network within their budget. The management of multiple service providers by a single organization presents significant administrative challenges.

SIAM provides a standardized methodology for integrating and managing multiple service providers and their services. It enhances the management of the end-to-end supply chain and provides governance, management, integration, assurance and coordination to maximize the value received.

SIAM supports cross-functional, cross-process and cross-provider integration in a complex sourcing environment or ecosystem. It ensures all parties understand and are empowered to fulfil their role and responsibilities, and are held accountable for the outcomes they support.

SIAM recommends the appointment of a single logical entity with accountability for end-to-end service delivery: the service integrator. The customer organization has a management relationship with the service integrator, and the service integrator manages the relationships with service providers.

1.3 History of the SIAM Foundation BoK

SIAM evolved from many different organizations and countries. As organizations developed proprietary materials, there was little objective guidance available for practitioners.

In 2016, in response to the requirement for SIAM guidance, Scopism Limited worked with contributors from a wide range of organizations and individuals to create the SIAM Foundation BoK. This publication provides the basis for the subsequent SIAM Foundation certification scheme launched by EXIN and Scopism.[2]

1.4 SIAM key concepts

The following sections describe SIAM key concepts. More detail is available in the **SIAM Foundation BoK**[3]:

- SIAM layers
- SIAM structures
- Drivers for SIAM
- SIAM terminology:
  - SIAM practice
  - SIAM function
  - SIAM roles
  - SIAM structural elements
  - SIAM models
- SIAM roadmap

1.5 SIAM layers

The SIAM ecosystem consists of three basic layers: the customer organization, the service integrator and service providers.

---

[2] For more information, visit: *www.exin.com/certifications/exin-siamtm-foundation-exam*.

[3] *www.itgovernancepublishing.co.uk/product/service-integration-and-management-siam-foundation-body-of-knowledge-bok-second-edition*.

The focus, activities and responsibilities for each layer are different, as shown in **Figure 1**.

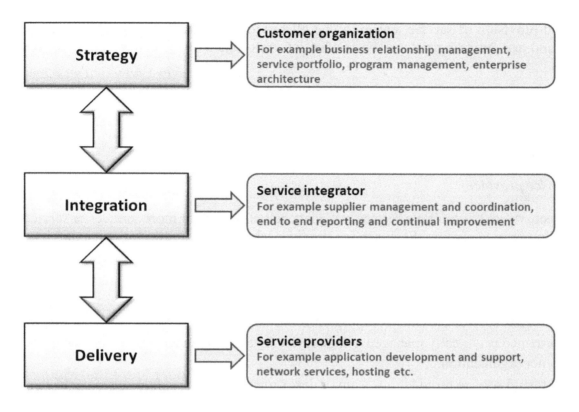

Figure 1: SIAM layers

An overview of each layer is provided here, with detailed information contained in the **SIAM Foundation BoK**.

1.5.1 Customer organization

The role of the customer organization is to commission services and provide direction based on the organization's strategy. In a traditional multi-service provider model, the customer organization has a direct relationship with each service provider. In a SIAM model, the customer organization has a relationship with the service integrator. It retains ownership of the commercial relationship with each service provider, however, the service integrator carries out management, governance, integration, coordination and assurance activities.

The customer organization may have 'retained capabilities', which are skills and capabilities important for the delivery of service. Retained capabilities may sometimes be referred to as the 'intelligent client function'.

1.5.2 Service integrator

The service integrator is responsible for managing service providers. It provides governance, management, integration, assurance and coordination across the SIAM ecosystem. It focuses on the end-to-end provision of service, ensuring that all service providers are properly engaged in service delivery and are providing value. The service integrator encourages collaboration between service providers.

The service integrator layer may be fulfilled by one or more organizations, including the customer organization. Having more than one organization in the service integrator role provides an additional challenge, so this approach must be managed carefully to ensure roles and responsibilities are clearly defined. (See section **1.6.3 Hybrid service integrator**.)

1.5.3 Service provider

A SIAM ecosystem has multiple service providers delivering one or more services or service elements to the customer organization. Each service provider takes responsibility for managing its part of the contracted service, including the technology and processes that support end-to-end service delivery.

Service providers can be part of the customer organization or external to it:

- An **external** service provider is an organization that is not part of the customer organization. Its performance is typically managed using service level agreements (SLAs) and a contract with the customer organization.
- An **internal** service provider is a team or department that is part of the customer organization. Its performance is typically managed using internal agreements and targets.

It can be helpful to categorize service providers according to their importance and potential impact on the customer organization, which will also indicate the level of governance required for each of them. The commonly used categories are strategic, tactical and commodity. SIAM applies to all three categories, but the nature of the relationship and the management time required will differ.

1.6 SIAM structures

There are four common SIAM structures, differentiated by the sourcing and configuration of the service integrator layer. These are:

1. Externally sourced service integrator
2. Internally sourced service integrator
3. Hybrid service integrator
4. Lead supplier as service integrator

The decision to select a structure will depend on factors including, but not limited to:

- Business requirements
- Internal capabilities
- Complexity of the customer's services
- Customer organization type and size
- Legislative and regulatory requirements

- Customer budget
- Existing service management capability in the customer organization
- Timescales
- Types and numbers of service providers in the ecosystem
- Customer organization maturity and risk appetite

1.6.1 Externally sourced service integrator

Figure 2 illustrates the externally sourced service integrator structure. An external organization is commissioned by the customer organization to act as the service integrator. The external service integrator is solely responsible for managing service providers and does not have any service provider responsibilities.

This structure is suitable for customers that do not have sufficient skills or capabilities to be a service integrator, do not want to develop them and are prepared to trust an external organization to fulfil the role.

There are distinct advantages and disadvantages to this structure, as discussed in the **SIAM Foundation BoK**.

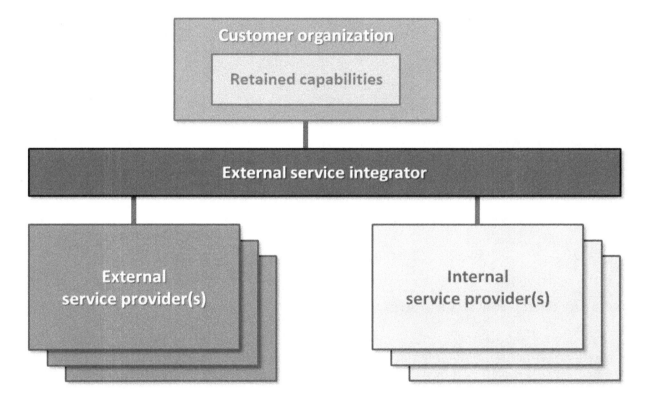

Figure 2: Externally sourced service integrator

1.6.2 Internally sourced service integrator

In this structure, the customer organization takes the role of service integrator. The service integrator must still be viewed as a separate, logical entity. If the roles of customer and service integrator are not separated, then the model is simply that of a traditional organization with multiple service providers, losing the benefits of SIAM.

As shown in **figure 3**, the service providers can be either internal or external.

This structure is applicable to customers that have, or wish to develop, capabilities in service integration. The advantages and disadvantages of this structure are detailed in the **SIAM Foundation BoK**.

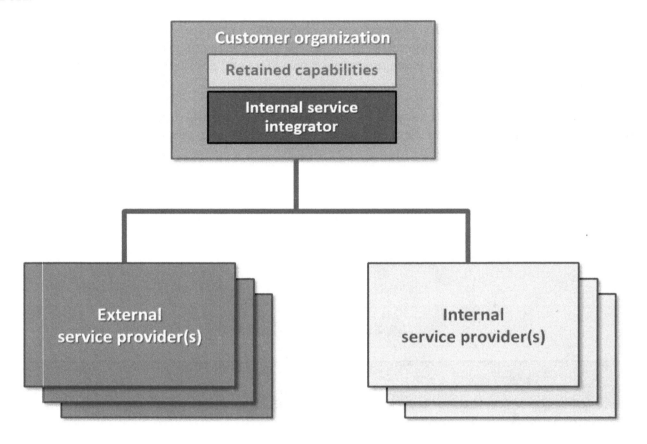

Figure 3: Internally sourced service integrator

Structuring the internal service integrator

Different SIAM models will place the internal service integrator in different parts of the organizational structure. Some internal service integrators reside in the IT department, others may be a separate department within the organizational structure. The structure will depend on several factors, including the scope of the SIAM model and the size of the organization.

1.6.3 Hybrid service integrator

In the hybrid structure, the customer collaborates with an external organization to provide the service integrator capability, as shown in **figure 4**. As with other structures, the service provider roles are carried out by internal or external providers.

This structure is suitable for customer organizations that wish to retain an element of control over the role of service integrator, but do not have the skills to perform all aspects. Some elements are provided internally from existing resources, while others are sourced externally. This model is useful when a customer organization wishes to develop service integrator skills and can draw on external expertise while acquiring them. This structure may be a temporary arrangement, until the customer organization has sufficient skills to carry out the service integrator role alone, or, until the customer organization migrates to a wholly external service integrator.

Although this structure is potentially complicated in terms of allocating roles and responsibilities, the advantages may outweigh the disadvantages in some situations. More information is contained in the **SIAM Foundation BoK**.

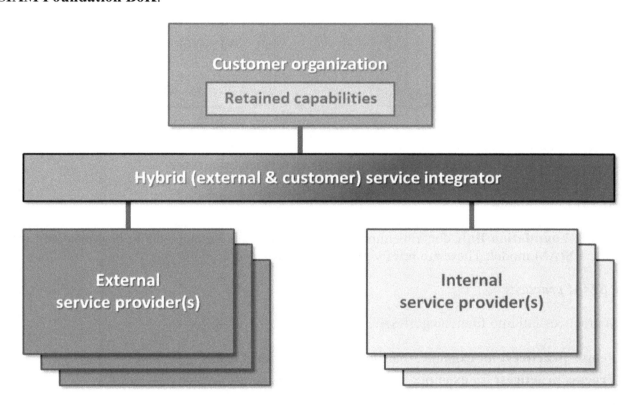

Figure 4: Hybrid service integrator

1.6.4 Lead supplier as service integrator

The lead supplier structure includes one organization acting in the role of service integrator as well as a service provider. This is illustrated in **figure 5**.

The reasons for selecting this structure are similar to those for selecting an external service integrator. It is suitable for customers who have an existing relationship with a service provider with integration capabilities (or service integrator with delivery capabilities). The advantages and disadvantages to this structure are similar to those of the externally sourced service integrator, but there are some additional considerations, as discussed in the **SIAM Foundation BoK**.

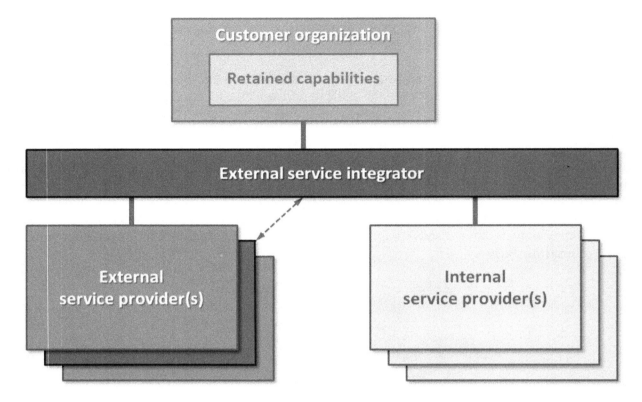

Figure 5: Lead supplier as service integrator

1.7 SIAM terminology

The **SIAM Foundation BoK** contains important information on the elements to be considered when adopting a SIAM model. These are briefly reviewed in this section.

1.7.1 SIAM practices

SIAM practices fall into four categories:

- **People practices:** for example, managing cross-functional teams
- **Process practices:** for example, integrating processes across service providers
- **Measurement practices:** for example, reporting on end-to-end services
- **Technology practices:** for example, creating a tooling strategy

1.7.2 SIAM functions

Each organization in a SIAM ecosystem will have its own structure, processes and practices. In each layer, there will be processes and practices that are specific to the role of the organization.

The service integrator layer includes functions relating to operational governance, management, assurance, integration and coordination. These will not be the same for the customer organization or the service providers. In each SIAM ecosystem, careful consideration must be given to the activities carried out by each organization and how they interact with other providers.

1.7.3 SIAM roles

Clearly defined roles and responsibilities ensure that a SIAM ecosystem will work effectively. A common cause of SIAM implementation failure occurs when roles and responsibilities have not been considered or fully understood.

Roles applicable to the different SIAM layers are defined and implemented as part of the SIAM roadmap. Each SIAM model will have its own specific requirements, and these need to be defined, established, monitored and improved. This includes the roles and responsibilities of each layer, organization, function and structural element.

1.7.4 SIAM structural elements

The term 'structural elements' refers to entities that have specific responsibilities working across multiple organizations and SIAM layers. These structural elements connect the functions from each layer to the processes, practices and roles across the SIAM ecosystem.

There are three types of structural elements:

- Boards
- Process forums
- Working groups

Structural elements include representatives from the service integrator, service providers and, where required, the customer organization. This encourages collaboration and communication across the ecosystem, so that all parties work together to achieve shared goals.

1.7.5 SIAM models

There is no single 'perfect' SIAM model. Each organization develops its own model based on its own requirements, the services in scope and desired outcomes. Organizations may draw on proprietary models provided by an externally sourced service integrator or external advisors and consultants engaged during the SIAM transformation. Whichever SIAM model is chosen by the customer organization, it will share common characteristics, as shown in **figure 6**.

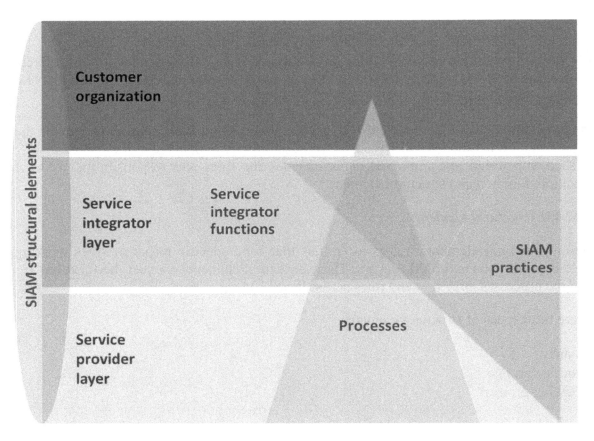

Figure 6: A high-level SIAM model

1.8 SIAM roadmap

The SIAM roadmap describes the high-level stages and activities required to create and transform to a SIAM model. It consists of four stages, shown here with their objectives and main outputs.

1. **Discovery & Strategy**: initiates the SIAM transformation project, formulates key strategies and maps the current situation.

Outputs include:

- An established SIAM transition project
- Strategic objectives
- Governance requirements and high-level SIAM governance framework
- Defined principles and policies for roles and responsibilities
- Map of existing services and sourcing environment
- Current maturity and capability levels
- Market awareness
- Approved outline business case for SIAM
- Strategy for SIAM
- Outline SIAM model

2. **Plan & Build**: completes the design for SIAM and creates the plans for transformation.

Outputs include:

- Full design of the SIAM model including:
  - Services, service groups and service providers (the 'service model')
  - The selected SIAM structure
  - Process models
  - Practices
  - Structural elements
  - Roles and responsibilities
  - Governance model
  - Performance management and reporting framework
  - Collaboration model
  - Tooling strategy
  - Ongoing improvement framework
- Approved business case
- Organizational change management (OCM) activities
- Service integrator appointed
- Service providers appointed
- Plan for service provider and service retirement

3. **Implement**: manages the transition from the current 'as-is' state to the 'to-be' SIAM model. The output from the Implement stage is the new operational SIAM model supported by appropriate contracts and agreements.

4. **Run & Improve:** manages the SIAM model, day-to-day service delivery, processes, teams and tools, and continual improvement.

Outputs from the Run & Improve stage fall into two categories:

- **Run outputs:** business as usual (BAU) outputs including reports, service data and process data.
- **Improve outputs:** information used to evolve and continually improve the SIAM model.

CHAPTER 2: SIAM ROADMAP STAGE 1: DISCOVERY & STRATEGY

The Discovery & Strategy stage analyzes the customer organization's current situation, formulates key strategies and initiates a SIAM transition program, if appropriate. This enables the customer organization to:

- Confirm whether SIAM is an appropriate approach, based on expected benefits and risks
- Determine a sourcing strategy, based on existing service providers; those that can be retained and activities suitable for external sourcing
- Consider additional skills and resources that may be required for the SIAM transition and subsequent operation of the SIAM ecosystem

---

Why SIAM?

In an IT environment where many services are becoming commoditized (cloud, as-a-service, etc.) and where multiple vendors need to work together to provide business-critical services, many organizations are spending more time on supplier management, rather than on actually delivering services. They are increasingly considering SIAM to:

- Understand the end-to-end picture of service provision
- Coordinate the activities of multiple service providers
- Provide a single source of truth regarding service performance
- Be a trusted partner in developing new services and strategies
- Optimize delivery through people, processes, tools and suppliers
- Ensure smooth performance of day-to-day operations, enabling them to concentrate on more progressive activities

---

2.1 Roadmap 'flow'

There is no one correct way of 'doing' SIAM. Most commissioning organizations already work with one or more service providers, and have different objectives, priorities and resources. Many elements influence the decision to adopt SIAM, along with deciding the appropriate approach for the transition to the chosen SIAM model.

Discovery & Strategy is a critical stage of the roadmap, as each customer organization's maturity, services and level of SIAM readiness is different. For example, some organizations may already have a sourcing strategy or mature supplier management capabilities, whereas others will need to create these as part of their SIAM roadmap. If activities are missed – or are partially completed – there could be a negative impact on the remainder of the transition project activities.

Many of the outputs from Discovery & Strategy are refined and expanded in the Plan & Build stage. A SIAM program is likely to require an iterative approach. Often, completion of a task will lead to the revision of an earlier activity. For example, designing the detailed SIAM model in the Plan & Build stage may lead to a review of the SIAM strategy, an activity completed in the Discovery & Strategy stage. It is essential to revisit the approach regularly and reassess previous decisions when needed.

The sequence of activities in the roadmap is not a predefined 'checklist' or prescriptive approach. Rather, it is an optimal approach, based on the authors' experience. Each element is addressed as part of a SIAM roadmap; each task's structure, order and priority depends on the customer organization's particular circumstances. Time pressures may require activities to happen in parallel for some organizations.

Designing without tailoring

The CIO of a small organization with 50 users created a SIAM strategy before understanding the capabilities of their internal IT team and its current services.

They engaged external consultants to design a SIAM model. They reused a model that they had created for a large, multinational organization. This included contract schedules for an external service integrator and several service providers. The CIO then brought in a separate team of external procurement professionals to run a major procurement exercise. After 12 months, the service integrator and service providers were appointed.

Because of the lack of tailoring to a 'standard' SIAM model to suit the needs of the smaller organization, the result was a threefold increase in the costs of the IT services.

2.2 Establish the SIAM transition project

"The SIAM transition project should be formally established using the organization's selected project management methodology."[4]

Managing the transition to a SIAM model is a significant undertaking. The time, costs, effort and resources for all parties involved should not be underestimated. This section provides some guidance on the most appropriate project management methodologies and approaches that can be used throughout the SIAM roadmap stages.

The context of this section is to provide guidance regarding the discovery, planning and implementation of a SIAM model. It is not intended to provide recommendations for the implementation of a project management framework.

2.2.1 Project management methods

There are a variety of project management standards, methodologies and frameworks available, including:

---

[4] *Service Integration and Management (SIAM™) Foundation Body of Knowledge (BoK), Second edition*, IT Governance Publishing, 2021. For more information visit: *www.itgovernancepublishing.co.uk/product/service-integration-and-management-siam-foundation-body-of-knowledge-bok-second-edition*.

- International standards, for example ISO 21500
- Country specific standards, for example ANSI, BS 6079, DIN 69900:2009
- Generic or global methodologies, for example PMI, PMBOK, APM, SCRUM, PRINCE2
- Industry specific practices, for example HOAI, V-Model, etc.

Combining these with any existing practices within an organization provides a good starting point for managing the SIAM transition project and the activities of the SIAM roadmap.

Minimize complexity

Most organizations will already have a preferred project management methodology. It is usually better to adopt that approach, rather than impose a different method that can lead to further disruption as the organization learns SIAM principles and a new project management approach.

Successful completion of complex projects, such as a SIAM transition, almost invariably requires the selection and expert application of a number of different enabling practices, approaches and frameworks. It is important to consider the mix and the level of expertise that will be required for a SIAM transformation.

Some organizations may choose to work with an external provider of project management capabilities. This may arise when the organization has no capability in project management, no available resources or little experience of managing a project of this scale and type. In these cases, it can be helpful to work with an organization that adheres to an organizational or global project management standard. This reduces the risk of becoming over-reliant on a proprietary framework and its provider.

Regardless of the project management methodologies and practices selected, it is important to reach agreement on the principles and approaches to be used. This ensures a common understanding between the stakeholders involved in project delivery and governance.

In many organizations, the transformation to a SIAM model may be carried out as a program with several projects within it. In this case, a program management method or approach will need to be considered, in addition to a project management methodology.

Program versus project

Although programs and projects have many similarities, they also display several different characteristics and functions.

A **project** is well defined, with a start and end point and specific objectives that, when attained, signify completion.

A **program** has greater levels of uncertainty. It can be defined as a group of related projects managed in a coordinated way to obtain benefits not available from managing projects individually.

The transition to SIAM is usually a program that can span many years and includes several discrete projects. During the Discovery & Strategy stage, there will be many unknowns and variables to discover, define and resolve. The outcome of this stage is an outline business case for a SIAM transition project, and for the remainder of the roadmap stages (see section **2.7 Create an outline business case**).

2.2.2 Agile or waterfall?

Agile (as detailed in the Agile manifesto[5]) is a set of methods and practices for software development. It is based on iterative and incremental development, and a rapid and flexible response to change. Requirements and solutions evolve through collaboration between self-organizing, cross-functional teams.

Although a SIAM project is not a software development project, Agile practices can add value. This topic is discussed in more detail within the **SIAM Foundation BoK**, including an alignment of Agile principles and how they might be applied to SIAM.

A waterfall approach is a more traditional method of development and implementation that follows sequential stages and a fixed plan of work. For example, plan, design, build and deploy. There are many activities within the SIAM program that can be addressed in an Agile or iterative manner, whereas other activities may be more appropriately coordinated in a traditional waterfall approach that promotes more detailed planning. A hybrid approach might use waterfall to set out the overall milestones, but apply iterations within the stages to achieve them.

Waterfall and Agile

At a UK manufacturing organization, the incoming service integrator created a waterfall implementation plan.

The transfer of processes was broken down into a number of phases and rolled out to the service providers being incorporated into the SIAM model. The processes were to be documented by the service integrator and introduced to service providers through a series of workshops.

However, during due diligence it became apparent that some key processes were suitable for a single service provider, but would not work in a multi-service-provider model. It was necessary to redevelop these processes as a priority, using an agile approach, in conjunction with the new service providers. This approach helped to ensure that the processes would apply to all service providers in the SIAM ecosystem. Other processes were not implemented in full, but were developed to a state that enabled them to be adapted and used much earlier than had been expected.

The result of this collaborative development approach was that the new service providers had much better commitment to the processes and a healthy working relationship emerged between the service integrator and service providers. The customer organization had the benefit of some processes that were implemented earlier than expected and with evidence that collaboration would work effectively.

2.2.3 Project governance

Project governance is the management framework within which project decisions are made. Project governance is separate from overall IT, organizational or SIAM governance, and provides a set of rules and regulations for all projects, irrespective of whether they are Agile or waterfall.

---

[5] *agilemanifesto.org*.

Project governance does not describe the governance of the SIAM ecosystem. As part of a SIAM transition project, a governance model for SIAM is created, as defined in section **2.3 Establish a SIAM governance framework**.

Governance is a critical element of any project. It provides a framework of accountabilities and responsibilities associated with an organization's projects. It is fundamental to ensuring control and contributing to the overall success of the project. The role of project governance is to provide a decision-making framework that is logical, robust and repeatable. It assures that decisions and directions occur in a correct and timely fashion. There are a variety of options available to support decision making within a project, including:

- Consensus decision making
- Majority vote
- Delegating the decision to an expert or subgroup, allowing the SIAM governance lead to make the decision after discussion (see section **2.3.7.1 SIAM governance lead**)

Governance focuses on the organization's requirements and should be business (not IT) oriented. Governance processes should be linked to business value and measured against them. Project governance is based on the overall strategy of the customer organization. Project managers must understand the customer organization's objectives and vision, and how these relate to the project governance framework.

The project governance framework provides a mechanism to maintain visibility of the project status and to understand and manage the risks associated with the project. The most desirable approach is to create a project governance framework that allows projects to achieve results unhindered by bureaucracy, micromanagement and unnecessary scrutiny.

2.2.3.1 Elements of project governance

Examples of project governance artefacts and elements include:

- Project mandate
- Project initiation document
- Business case
- Business requirements
- Agreed specification for the project deliverables
- Critical success factors (CSFs)
- Clear assignment of project roles and responsibilities
- Project steering committee/board and project manager
- Stakeholder map identifying all stakeholders with an interest in the project, classified by their role and influence
- Communication plan – defining the method of communication to each type of stakeholder
- Organizational change management (OCM) approach
- Project and/or stage plans
- Resource requirements
- System of accurate status and progress reporting
- Method for comparing the completed project to its original objectives

- Central document repository for the project
- Centrally held glossary of project terms
- Process for issue management
- Risk and issue log – and a process for recording and communicating these during the project
- Standard for quality review of the key governance documents and of the project deliverables

This list may not be necessary for every SIAM project. Each one needs to consider the artefacts that are required for a successful outcome.

2.2.3.2 Project roles

The organizational structure of a SIAM transition project needs to be resilient so it will remain stable throughout the entire project. Changing the project roles and responsibilities midway is disruptive and should be avoided as far as possible. Individuals fulfil roles, and although individuals can change, the roles and responsibilities within the project structure remain constant.

There should be a project team that represents the service integrator (once appointed) and representatives from the key incumbent service providers. Representation from the customer organization is also needed, especially during the early stages (Discovery & Strategy, and Plan & Build) when decisions about the SIAM model are made. As additional organizations are appointed and onboarded (for example, an external service integrator or service providers), they must be included within the project structure.

Some customer organizations choose to delegate their role in the project structure to the service integrator (once appointed), particularly if the service integrator is sourced internally. In an external service integrator structure, the customer organization may want to maintain representation in the project team to avoid becoming detached from project management and decision-making activities. Its role in the project structure is to give direction, provide oversight and be seen by the stakeholders and parties involved as supportive of, and committed to, the SIAM transition project.

Representation within the project

Not every service provider will be interested in becoming part of the project team. The level of engagement varies, depending on the service provider's delivery model. The SIAM transition approach should be flexible enough to accommodate this.

For example, a large cloud solution provider will not, in most cases, send representatives to its customer's project management team or meetings.

Organizations must ensure that if an external service provider is unable or unwilling to become part of the project structure (perhaps because it has not yet been appointed), there remains representation for that role, to provide input and consideration to project management. An example would be a service owner from the retained capabilities that can represent the external service provider for that service within the project structure.

Independent of the applied project management approach and governance framework, there are several structural elements that have been demonstrated to be effective in SIAM transition projects and programs.

Project board

> **Project board**
>
> The project board is responsible for overseeing and steering the SIAM transition project, making project decisions, adjusting the project when necessary, and providing communication and guidance throughout the transition period. It is sometimes referred to as a steering committee.

Having a functioning project board involving all relevant stakeholders is essential for effective project management and swift decision making. The project board steps in when a lower level in the project structure, for example, the project manager, needs higher authority for decisions or where they cannot resolve an issue, for example, a lack of resources.

PRINCE2 provides a definition of the roles for a project board that can be applied to a SIAM transition project:

- **Executive:** sponsor from the customer organization who ensures that the program/project meets expected benefits and is adequately equipped with resources.
- **Senior User:** someone with adequate authority to represent the consumers of a SIAM program outcome, typically from the retained capabilities.
- **Senior Suppliers:** representatives from the various service provider organizations, including the internal and/or external service integrator.

> **Senior User**
>
> A service provider may also act as a Senior User, for example, for tools that they will be required to use, such as the incident management system. If its tools are being replaced, it may be useful for the service provider to help define and test the quality criteria for the new ones.

The project manager manages a project on a day-to-day basis on behalf of the project board, within specified constraints. The customer organization may provide a project manager, source a project manager for the SIAM transition project or ask the service integrator (once appointed) to fulfil this role.

Transition Review Board

A Transition Review Board is a joint board of service providers, the service integrator and the customer's retained capabilities. It is an optional board with a role that is more operational and 'hands on' than the project board, and has a place between them and the project manager. This board can be useful later during the Plan & Build, and Implement stages, when the service integrator and many of the service providers have been appointed.

This board is responsible for:

- Reviewing transition progress against planned activities
- Reporting progress to the project board
- Identifying and managing project issues

- When necessary, escalating issues or decisions to the project board
- Ensuring that all parties are aligned in terms of plan execution
- Making the recommendation to the project board to accept the outcomes of one stage and commence the next, including final acceptance of the implementation and handover to operations

Once the project is completed, this board may be retired, or alternatively continue as a board within the SIAM governance framework, for example the Service Review Board (see section **2.3.7.4 Governance boards**).

Project Management Office

A Project Management Office (PMO) is a group or department that defines and maintains standards for project management within the organization. The PMO strives to standardize and introduce economies of repetition in the execution of projects. The PMO is the source of documentation, guidance and metrics on the practice and execution of project management.

The PMO may have other functions beyond standards and methodology, and may participate in strategic project management, either as facilitator or as owner of the service portfolio management process within the SIAM model. It is responsible for monitoring and reporting on active projects and portfolios, and reporting progress to senior management for strategic decisions. A PMO can provide the following benefits during a SIAM project:

- Align the portfolio of projects, including the SIAM project, with other activities within the customer organization
- Provide advice on lessons learned in previous projects that can help to avoid mistakes and incorrect approaches
- Monitor and report on the progress of projects, whether they are on time and within budget, according to the defined scope, along with tracking risk and issues
- Understand the links and dependencies between projects, for example, multiple projects within a SIAM program
- Improve communication within the program and project teams and stakeholders, including service providers, the service integrator and others
- The customer organization's PMO can provide guidance to the service integrator and service providers involved with the project to improve standardization within the SIAM environment
- Provide or assure document control and maintenance of a document repository for projects

Solution architecture and assurance function(s)

Different stakeholders are involved in the strategy, planning, building and implementation of the SIAM transition project. To support these stages, a detailed SIAM ecosystem is designed, including a process model, tooling strategy and reporting framework.

Designing a SIAM model requires specialized capabilities. Forming a solution architecture team will allow for the inclusion of stakeholders, including subject matter experts (SMEs), such as enterprise architects. Integrated architecture principles can be applied to the design of any SIAM model, as they provide a strong focus on business requirements and drivers. The need for all aspects of the architecture and all architectural decisions to be traceable back to business priorities must always be considered.

Enterprise architecture frameworks

Most enterprise architecture (EA) frameworks divide the architecture description into domains, layers or views, and offer models – typically matrices and diagrams – for documenting each view.

This allows for systemic design decisions on all the components of the system and long-term decisions around new design requirements, sustainability and support.

In addition to a solution architecture function, it is beneficial to introduce a solution assurance function early and for the life of the project. Often, early design decisions must be revisited or amended to comply with the agreed approach and principles. Sometimes, solution assurance and architecture functions are combined into a single team. Alternatively, the solution assurance team can be incorporated within the PMO, so it can provide assurance across multiple projects.

Communications team

Communication is a CSF for any organizational change program. It is important to plan, structure and design communications to support change towards a SIAM culture. The dedicated communications team will be part of the project structure, spanning the affected organizations. Communication activities across the layers include:

1. **Customer organization:** for large organizations that operate in different geographies, regional and country stakeholders typically consume and require most communication efforts. Different forms of resistance to change can be observed within these stakeholder groups in almost all SIAM projects, and these objections need to be anticipated and addressed.
2. **Service integrator:** needs to execute and supervise communication activities from a central communication team's perspective to its own personnel, as well as to the service providers in the SIAM model.
3. **Service providers:** ideally, communication professionals in the service provider layer should form an extension of the service integrator communications team for execution and feedback.

Procurement and contract management

Many SIAM transition projects include the selection of new service providers, and all include the creation of new contracts and the retirement or termination of existing contracts. It is important to establish structures and plans within the project for these activities.

The contract management team needs to assess existing contracts, review and learn from existing issues, understand the target SIAM model and establish appropriate new contracts. The contract management team should, therefore, be embedded into the SIAM transition project structure. If not, there is a risk that working in isolation will result in disparities between the contractual scope and what is required for the chosen SIAM model.

The contract management team can be established from within the customer organization's retained capabilities or provided externally, if the internal capability is not sufficiently mature. Eventually, accountability for contract management for the SIAM model must reside within the customer's retained capabilities. In situations where the initial contract management team was sourced from an external organization, a plan to form a contract management team within the customer organization must be delivered as part of the SIAM transition project.

The procurement management team should agree its approach with the project team, in line with the strategic objectives and the chosen sourcing approach (see section **3.1.2 Sourcing approach and the selected SIAM structure**). It also needs to prepare procurement documentation, using artefacts from the SIAM project team and contract management teams, ensuring that it complies with legal requirements. It will then manage the procurement process. This may include dialogue with the service providers to elicit requirements that need their input or to address situations where providers are unwilling to accept particular responsibilities.

The procurement management team will conduct negotiations with potential service providers until the contracts are signed. During these negotiations, it must maintain alignment with the SIAM strategy and model. Many customer organizations use external procurement resources as they rarely have the capability or experience in sourcing providers for a SIAM ecosystem. If external resources are used, consider how this capability will be provided once the SIAM ecosystem is established.

The activities for contract management and procurement management must be embedded in the overall plan for the SIAM transition (see section **3.3.3 Transition planning**) as part of the critical path to success.

2.2.3.3 Building a 'one team' culture

It is essential that the Discovery & Strategy stage defines the behavioral expectations from all stakeholders. Building a 'one team' culture should start early with the project team. Culture can make a major difference to the success of SIAM activities in later roadmap stages so must be considered in detail at every stage of the roadmap.

Building trust and acting as a single team across multiple organizations can be a challenge and is discussed elsewhere in this guidance in relation to culture and behavior (see section **3.2 Organizational change management approach**).

The way in which project governance is conducted will have a significant impact on the trust and behaviors instilled in the SIAM environment. The behavior of project team members and their engagement with others will set a benchmark for the service providers as they join. When managing the project, it is important to pay attention to 'fairness' and whether stakeholders feel their opinions are being heard. 'Being heard' may require a new approach to prevent some stakeholders dominating meetings.

For example, if the service integrator exhibits the desired behavior, the service providers are more likely to display the same. If a service provider feels that another service provider is being favored, it may avoid the service integrator and escalate the issue to the customer's retained capabilities, causing project delays.

It must be clear to all service providers from the outset that they will be expected to work with other service providers on the project, and to share information about progress and issues. This can be supported by techniques such as the Chatham House Rule, where participants are free to use information received to progress the SIAM project, but not to disclose the source of information or to use it outside the SIAM project.

Chatham House Rule

The Chatham House Rule:

"When a meeting, or part thereof, is held under the Chatham House Rule, participants are free to use the information received, but neither the identity nor the affiliation of the speaker(s), nor that of any other participant, may be revealed." [6]

The rule originated at Chatham House with the aim of providing anonymity to speakers and to encourage openness and the sharing of information. It is now used throughout the world as an aid to free discussion.

Collaboration and openness are crucial to the success of a SIAM model. The commitment of competing service providers to work together should be tested during procurement, to give an indication of whether service providers will be a good cultural fit.

Many commercial sensitivities need to be recognized and overcome if the project is to be successful, so the governance model for the project must address these from the outset. It is advisable to establish a cross-service provider body that focuses on partnership and collaboration as early as possible in the SIAM project (see section **2.3.7.4 Governance boards**, Partnership and Collaboration Steering Board).

A 'kick off' event focusing on delivering the SIAM project, attended by the customer, retained capabilities, service integrator and service providers (incumbent, or if already assigned) can be a useful first step in creating the common vision necessary for the 'one team' culture and building trust.

2.2.4 Implementation approach

It is very important to set the right expectations with senior stakeholders when creating initial project plans and securing sufficient budget for the roadmap stages. Sometimes the cost of SIAM projects is understated, the benefits are overstated or the timelines shortened to ease business-case approval. If the business case is unrealistic, there is a high risk that the customer organization will see the project as a failure because its expectations are not met.

Project management methodologies allow a project to be divided into discrete stages. Provided that the complete transformation scope has been signed off at the 'initiation stage' of the project, the detail of these stages can be determined later, preferably without the need to seek more funding. As suggested in section **2.2.2 Agile or waterfall?**, an iterative Agile approach provides a phased implementation, with each stage learning and adapting from previous stages.

Iterations of integration – a process example

Iterative approaches must still deliver results. For processes, the important step is to establish the roles and responsibilities, and interactions and activities to support them. The first iteration for a

---

[6] Royal Institute of International Affairs: *www.chathamhouse.org/about/chatham-house-rule.*

> process might be manual integration activities (for example, a 'swivel chair' approach for logging incident records).[7] Then, the next iterations leading to data integration can be developed and implemented based on feedback and lessons learned.

Before you start

Any journey requires a good understanding of the objectives and outcomes – the 'destination', way points, and the starting point or 'point of origin'. Discovery and assessment of the current state is essential. This includes baselining critical resources for the transition, including their capability, maturity, availability or any other constraints (see section **2.5 Analyze the current state**).

The chosen project management approach should help set standards for project artefacts and stages, such as a business case, setting up and initiating the project. These early stages are important, because they set senior management expectations (and commitment) for the project. Those expectations, for example regarding timing, budget and quality outcomes, are not easy to change during later stages. Independent of the benefits actually achieved from the SIAM model, setting the right expectations will often determine the perceived success or failure of the project.

2.2.5 Outcomes, objectives and benefits management

The project outcomes and objectives should be aligned to the strategic objectives defined by the customer organization. Quality and acceptance criteria should be defined for the project as part of a benefits map or profile, and within the project's product descriptions and quality plans.

2.2.5.1 Distinction between transition and transformation

> **Transition versus transformation**
>
> **Transition** is the process of change from one form, state, style or place to another. A SIAM transition takes the customer organization from the previous, non-SIAM state to the start of the Run & Improve stage. Transition is a defined project with a start and end point.
>
> **Transformation** is the act of transforming or the state of being transformed. It is the overall SIAM program that makes all changes, in all roadmap stages, to realize the full benefits of the SIAM model. The customer organization is transformed from its previous way of working. Organizations need to complete a transition in order to transform how they operate.

2.2.5.2 Benefits realization management

Benefits realization management provides the customer organization with a way to measure how a SIAM project adds value to the enterprise.

Benefits realization approaches are defined and embedded into many project, program and portfolio management methodologies, and cover the full lifecycle from initial idea to realization. To assess if benefits have been delivered, the scope of the SIAM project must define a point when the implementation (or an implementation phase) is complete. As SIAM models evolve and improve

---

[7] For more information, visit: *www.webopedia.com/definitions/swivel-chair-interface/*.

continually, setting this point can be challenging, but it is essential when defining how benefits and success are to be measured.

Often, it is difficult to quantify project benefits and link these to the real outcomes and objectives for a SIAM transition (see section **3.5.2.4 Choosing the right measurements**).

Qualitative benefits

In reality, SIAM benefits are often based on subjective indicators such as a reduction in 'noise' to the customer organization, including its awareness of service provider disagreements or a general feeling that things are more under control.

Although benefits of this type are more difficult to measure quantitatively, in terms of value to the customer, they are important and must be considered as part of the performance framework.

Defining expected benefits (and how and when they will be measured) supports decision making during all stages of the project. It ensures that all decisions (particularly in relation to changes) are made with the benefits in mind. Often, project managers and other project team members become too focused on the detail of the project, forgetting objectives and required outcomes. This can lead to decisions that inadvertently affect the expected benefits adversely.

A defined set of objectives and an agreed way of measuring benefits, both financial and non-financial, is important. A clear and well-communicated business case, supported by ongoing project reporting, measurement and communication, will help to ensure appropriate expectation management.

Within Discovery & Strategy (as part of the outline business case), and Plan & Build (as detail is added to the business case), expected benefits need to be identified. Measuring SIAM benefits can be challenging, so this area requires a robust approach. The business case will clarify the purpose of creating and implementing a SIAM model, and its expected outcomes and benefits. The business case is a reference document that is used to clarify direction and purpose during the latter stages of the SIAM roadmap.

Benefits realisation plan

The Project Management Institute suggests that a benefits realization plan should:

- **Identify benefits** – ensure that the customer organization can identify expected benefits, to identify the best project and program investments
- **Execute/provide benefits** – look at practices that enable organizations to capture and realize both intended and unintended benefits
- **Sustain benefits** – provide practices that enable organizations to sustain benefits and achieve strategic objectives, to deliver ongoing value from outputs and outcomes once they transition back to the business

---

The benefits realization plan for a SIAM transition program or project should be created in the early roadmap stages. Checkpoints should be planned – both during the transition and beyond – once the SIAM model is deployed. Many benefits may not be fully realized until the SIAM model has been in place for a period of time. Checkpoints ensure all activities are on schedule to deliver planned benefits at the expected time. Where this is not the case, corrective action needs to be considered to address any shortfalls and avoid affecting the return on investment (ROI) planned in the business case. Effective planning for business benefits realization is key to evaluating the success of a SIAM model.

2.2.5.3 Expectations management

It is important to manage stakeholder expectations regarding the new SIAM model proactively. Good expectation management will reduce surprises and can identify a requirement to adjust the project course. These activities will support expectations management

- Goal setting and review
- Creating project plans
- Reducing unverified assumptions
- Openly communicating with all stakeholders
- Publishing status reports and project key performance indicator (KPI) information
- Avoiding unachievable commitments, or saying 'no' appropriately

Goal setting

To manage a successful SIAM transition, set mutually agreed goals across the SIAM layers that align to the customer organization's goals. When the customer organization's goals are clear, they can be used to assess changes. Will the change help the customer organization to reach its goals? Or is this simply a distraction? The service integrator will refer to these goals during conversations with stakeholders.

Creating project plans

Creating a project plan is necessary for a SIAM transition. Even if the customer organization claims not to care how the model is implemented or how activities are delivered, a plan with timelines can help set expectations and identify any assumptions or areas where more clarity is needed. The customer organization should always be able to understand the status of a project.

A SIAM transition requires the involvement and coordination of many different stakeholders. Adequate lead time is required to ensure that all stakeholders (both internal and external) can plan and mobilize appropriate resources required for the project.

Lack of planning

A large government organization started its SIAM transition by appointing an external service integrator. The service integrator invited all existing service providers to assist in process definition via a series of workshops. There was no overall plan or published timetable describing when SMEs would be required. The result was a series of workshops with 30–40 people all trying to agree on the correct models and how the processes should work.

Had the service integrator developed a plan first, defining whose input was required, when their attendance was needed and the activities to be performed, there would have been a better understanding of the purpose, objectives and required outcomes for the workshops. This approach would improve the chances of success in developing processes that are effective and fit for purpose.

Reducing unverified assumptions

Assuming that all stakeholders have the same understanding of a situation, project, deadline or task, presents risks to the success of the SIAM project. This issue can be easily overcome through adequate communication to discuss what is expected, how it might be accomplished and how benefits will be measured. One of the more common causes of project failure is the miscommunication of deadlines, potentially resulting in confusion and delays.

Openly communicate with all stakeholders

One of the best ways to manage expectations is to communicate frequently and clearly with all stakeholders within the SIAM project. Important times include the early stages of a new project and when a milestone or deadline approaches. It is important to strike a balance between sufficient and too much communication (sometimes referred to as communication overload). During a SIAM transition, service providers and stakeholders may be asked to work together for the first time. Ongoing communication can help to establish trust.

'Checking in' with the stakeholders throughout the course of the project makes it possible to provide real-time status updates and manage any delays, risks or bottlenecks. Proactive, honest and transparent communication fosters trust and introduces flexibility for making changes as the project proceeds. Being honest about a delay is better than promising to deliver and then missing a deadline. This is an essential part of SIAM culture.

Avoiding unachievable commitments

A large part of managing expectations is assessing if the expectation is appropriate.

Under promise, over deliver

In such complex environments, with so many stakeholders from different organizations and teams, it can be hard to give guarantees. There are too many factors that can affect what is delivered.

'Under promise, over deliver' is an old adage. It means that it is better to promise what you are certain of and do better if possible. Promising more than you are certain of creates a risk of failure. All the stakeholders in a SIAM model need to work together to define what can be promised realistically to the customer organization.

Expectations should be realistic and achievable. If they are not, it is acceptable to say 'no'. Being open with service customers and consumers regarding what can be delivered and, perhaps, what the plan to deliver other requirements at a later stage is, can contribute to instilling confidence and building effective long-term relationships.

2.3 Establish a SIAM governance framework

This activity defines governance requirements and creates a high-level governance framework to guide the future SIAM model. This is separate from project governance, defined earlier within section **2.2.3 Project governance**. The governance framework is used throughout all stages of the SIAM transition and operation (see sections **3.1.5 Governance model** in Plan & Build and **5.1 Operate governance structural elements** in Run & Improve).

What is governance?

Governance refers to the rules, policies and processes (and, in some cases, legislation) by which businesses are operated, regulated and controlled. There may be many layers of governance within a business, including enterprise, corporate and IT. In a SIAM ecosystem, governance refers to the definition and application of policies and standards across the SIAM layers. These define and ensure the required levels of authority, decision making and accountability.

Effective governance provides control and assurance that the SIAM ecosystem is properly aligned to the customer organization's requirements, is responsive to changes to those requirements and operates in accordance with strategy and plans. It also enables timely decision making, discussion of risks, issues and successes, and allows the views of different parties involved in service delivery to be heard by the customer organization.

2.3.1 IT governance

Effective IT governance addresses the challenges of IT provision. IT is a key capability and enabler for many organizations and is subject to a wide range of threats. Reliance on IT in today's interconnected 'always on' cyber landscape, and the immediacy of social media and 24/7 news coverage, creates the potential for significant financial and reputational impact caused by IT failures.

There are many IT options available to the customer organization, from 'as a service' to the cloud. Although these options can provide a faster and 'easier' route to a working solution, the overall complexity of the service landscape often increases because of an expanding portfolio of service providers, architectures and ways of working. Coupled with this is the increased potential for business units to bypass centralized IT governance and procure solutions themselves, potentially creating vulnerabilities and generating 'shadow IT' that the IT department is then expected to support.

IT governance involves organizations having appropriate structures and processes in place to enable management to make timely decisions, understand the risks that the organization faces and take steps to monitor and control those risks appropriately. Increasingly, legislation means that this is not just an IT problem, with large fines and board-level prison sentences potentially resulting from a failure to demonstrate adequate internal controls.

The customer organization needs a well-defined and robust approach to IT governance for internally and externally sourced services. The customer organization remains accountable for ensuring that risks are assessed and that appropriate controls have been implemented and are functioning correctly. This accountability cannot be outsourced, even if the responsibility for operation of some or all of the controls has been assigned to a service integrator.

2.3.2 Governance in a SIAM ecosystem

Within a SIAM ecosystem, governance ensures that there is a system of control in place across all the SIAM layers. Decisions are made considering all parties within the ecosystem and risks are clearly understood, controlled and monitored. The consequences of poor governance practices within a SIAM ecosystem can include:

- Inappropriate and sub-optimal allocation of contracts
- Breakdown of trust and relationships
- Lack of coordination when dealing with major incidents, problems and changes
- Charging based on 'what can be got away with'
- Delays because of poor communication and poor sharing of information
- Disputes related to unclear roles and responsibilities

In a SIAM ecosystem, governance and management operates at three layers:

- Strategic
- Tactical
- Operational

Governance practices involve:

- Clear accountability
- Fairness to all parties
- Ethical practices
- Openness and transparency
- Procedures to prevent conflicts of interest

These concepts fit well within the SIAM environment where there is a desire to encourage collaboration and partnership between the different organizations involved. At the highest level, SIAM governance focuses on three key aspects:

1. Ensuring alignment between the customer organization's current and future business needs and the SIAM strategy
2. Ensuring that the SIAM strategy and SIAM model are planned and implemented successfully
3. Ensuring that the SIAM model is managed, operated and improved in a controlled and collaborative manner, in compliance with both internal policies and external legislation

Within SIAM, boards are structural elements that perform a key role in providing governance and overseeing these aspects. They function as decision-making bodies, with representation from the appropriate level of the relevant stakeholders within the ecosystem.

2.3.3 Governance enablers

Guidance on effective governance can be obtained from established sources such as ISACA's COBIT governance and management framework for information and related technology. One of COBIT's guiding principles is that for efficient and effective governance, organizations need to take a holistic approach, focusing on seven categories of enablers that interact with and support each other as shown in **figure 7**. Each of these enablers needs to be considered when implementing governance within a SIAM context.

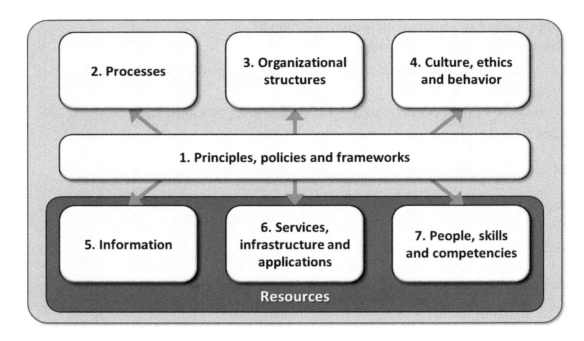

Figure 7: Enterprise enablers (ISACA COBIT framework)

Applying these categories of enablers to the SIAM model aids the identification of key components that need to be part of an organization's SIAM governance framework.

Principles, policies and frameworks are the steering mechanisms that guide the ongoing management and operation of the SIAM model. When properly constructed, these components inform decisions, underpin activities and ensure alignment across the ecosystem. Over time, requirements will change, so these enablers must be reviewed periodically and, if necessary, updated to maintain alignment with the customer organization.

Processes describe how particular objectives are achieved within the SIAM ecosystem, ideally aimed at ensuring an optimal mix of standardization and interoperability. Defined processes provide clarity about how things should work and enable repeatability, which, in turn, helps support desirable behaviors such as continual improvement. At the same time, within a SIAM environment, it is important to not enforce on the service providers the 'how' of process activities, but instead to focus on the 'what' (the outcome) and the interactions between different providers and with the service integrator (see section **3.1.3 Process models**).

Organizational structures provide decision-making bodies with defined scopes, roles and responsibilities. Within a SIAM ecosystem, a structure of governance boards, each with defined responsibilities, ensures that decisions are made at the appropriate level and with input from relevant parties, roles and skill sets.

Culture, ethics and behaviors. The importance of considering organizational cultures and the need to demonstrate ethical practices and embed key behaviors such as fair play, cannot be understated with regard to SIAM. Some organizations may be moving from a single provider model, where negative behaviors such as an 'us and them' mindset have become established. In other cases, there can be a significant difference in culture, ethics and behaviors between different service providers.

> **Individual versus group culture**
>
> It is sometimes the case that the individuals working within the SIAM ecosystem are honest, open and reasonable, but the group or corporate culture of their individual organizations is less so. The reverse can also be true. It is important to recognize that culture is often set at the top of an organization, so cultural issues may need to be escalated before the change can flow down to an operational level.

Transforming organizational culture to one of partnership, openness and fair play can be a significant challenge and will not happen by accident. Culture change needs to be planned and worked at, with clear leadership cascading down through the management chain. It may be necessary to move people into new roles if the incumbents are unable or unwilling to accept the change.

Information is a vital resource to support effective governance. Without accurate information about what is happening, how things are working and any issues being faced, it is impossible to make timely and informed decisions.

Services, infrastructure and applications are resources that enable the SIAM ecosystem to function efficiently and effectively. In a SIAM context, it is especially important to address interoperability as part of the tooling strategy and to ensure compliance with defined technical standards and data models.

People, skills and competencies are other important considerations within the SIAM ecosystem that need to be planned, monitored, managed and governed. Effective governance should involve regular review of whether the organization has access to the correct skills, confirmation that the people performing roles or making decisions have the necessary competencies and whether headcount requirements are changing.

Enabler categories

These enabler categories benefit from interaction with each other. For example, processes require people with the correct competencies to operate them, policies provide the 'rules' by which processes are operated, and applications store and allow the manipulation of information to support decision making. It is essential for organizations designing a SIAM model to review governance and management processes. Tools such as a RACI (Responsible, Accountable, Consulted, Informed) matrix can be used to show at which level of the planned SIAM model each process activity resides.

2.3.4 Governance requirements

To understand governance requirements, first consider what needs to be governed (the 'assets'), then identify and assess the risks that apply to those assets, before finally determining what controls, if any, need to be designed and implemented to manage those risks and provide assurance.

In the following sections, examples of assets and risks within a SIAM model are described at strategic, tactical and operational levels. These lists are not intended to be exhaustive, and much will depend on the specific organizations involved and the SIAM model selected. The intention is to provide common examples.

2.3.4.1 Strategic level

At a strategic level, examples of where governance may be required in a SIAM ecosystem include:

- SIAM business case (outline and full) and subsequent benefits realization
- Strategic plans
- Strategic risks and controls
- SIAM strategy (implementation and maintenance)
- SIAM model, including the SIAM process architecture
- SIAM tooling strategy and architecture
- SIAM organizational responsibilities
- Conformity to applicable external factors (for example, laws and regulations)
- Conformity to applicable internal factors (for example, organizational policies and standards)
- Ensuring alignment with corporate governance requirements, including any concerned with sustainability or environmental considerations

Once a list of the areas and items requiring governance has been identified, the next step is to identify the related potential risks (see section **2.3.11 Risk management**). In other words, 'What could go wrong?'.

Examples of potential risks include:

- Expected benefits from the SIAM program not being realized
- SIAM strategy and model failing to be wholly and properly implemented
- One or more parties not fulfilling their organizational responsibilities
- Unauthorized changes being made to strategic planning documents
- Sustainability or environmental obligations and targets not being met because of the actions of service providers within the SIAM ecosystem

Although in many cases it may not be necessary to carry out a full quantitative risk analysis, consideration should be given to identifying significant risks. Risks identified should be reviewed against agreed thresholds. The available resources, likelihood, potential impact and mitigation feasibility should be considered as part of decision making and risk prioritization. Attention then needs to be focused on the risks that are believed to be above the risk appetite threshold and how these might be managed, often using combinations of the governance enablers described earlier in this section.

Tracking benefits

Part of strategic governance should be to ensure that achievements are monitored and tracked against the original business case objectives (see section **2.2.5.2 Benefits realization management**). There should be regular reviews of whether the organization is still on track to achieve the SIAM program's strategic objectives and whether the intended value is being realized. The objectives will vary depending on the organization. For example, if one of the key objectives of the program was to increase business satisfaction with IT delivery, has this been achieved? If the objective was to obtain $4M cost savings over five years, is the program still on track to realize this?

In some organizations, there may already be mechanisms in place to track program benefits and value realized against business case objectives. Whether this is the case or not, the SIAM transition project/program management need to be aware of this information and track it. During the transition there will often be changes in scope, business requirements or external factors that could potentially have an impact on or alter the expected benefits. When this occurs, formal mechanisms are needed to approve the change (see section **2.2.3.2 Project roles**).

Part of this approval includes understanding and agreeing any changes to expected benefits. If the scope will be altered in such a way that the original objectives or value can no longer be achieved, then revised value objectives need to be agreed as part of change approval. It may be that management decides to reject the proposed change once they understand the impact it will have on value realization.

At all times, alignment must be maintained between the SIAM transition and its intended benefits, so that targets are attainable and success can be quantified and verified. If, at any stage, benefits tracking indicates that the expected benefits may not be realized, this will be reviewed and, where necessary, corrective actions implemented to bring the program back on track.

Governance related to the tooling strategy

Each SIAM transition project must consider how to deal with the implementation and configuration of a service management toolset. This is considered extensively in the Plan & Build section (see section **3.1.9 Tooling strategy**). The tooling strategy needs to be part of the governance considerations. Governance must ensure that there is a comprehensive, documented tooling strategy that includes the tooling ownership.

2.3.4.2 Tactical level

At a tactical level, examples of where governance may be required in a SIAM ecosystem include:

- SIAM plans (implement and operate plans for different process areas)
- SIAM tooling plans (implement and operate plans for different toolsets)
- SIAM process and tool ownership
- Service management data (ongoing ownership and governance)
- Ongoing initiatives, goals and plans, including service improvement plans
- Service provider transition plans (onboarding, handovers and offboarding)
- Management and coordination of projects
- Training and education plans
- Audit policy, plan and schedule

Governance in these areas helps create an understanding of the potential risks.

Examples include:

- Plans not being wholly or properly implemented
- Lack of ownership of plans, processes or tools
- Loss of valuable service management data, for example, during a change of service integrator
- Loss of key knowledge when service providers change or staff leave the organization
- Skill levels diminishing as new technologies are introduced
- Lack of coordination in multi-service provider projects, resulting in costly delays

The SIAM governance framework should include controls aimed at mitigating the risks at the tactical level.

2.3.4.3 Operational level

At an operational level, examples of where governance may be required in a SIAM ecosystem include:

- SIAM model
- Contracts and collaboration agreements
- Processes and procedures
- Work instructions
- Capabilities, skills and knowledge
- Toolsets
- Integrations and interfaces
- Policies and controls
- Process-related roles and responsibilities
- Access to and use of service management data
- Toolset configurations and associated documentation
- Defined KPIs and metrics covering:
  o Service integrator
  o Service providers
  o Processes and structural elements
- Operational reporting tools, templates and procedures
- Operational risks and controls
- Dispute resolution

Examples of common operational risks include:

- Lack of communication from higher governance levels causing misunderstanding and lack of buy-in for the new strategic direction
- Lack of understanding of the SIAM strategy by tactical and operational levels, leading to expected benefits of the SIAM program not being realized as actual operation does not comply with the intended strategy
- Contractual obligations not being fully understood or not being met
- Lack of collaboration and partnership
- Bypassing or lack of compliance with processes
- Interfaces between toolsets and/or processes being 'broken' because of unilateral changes
- Toolsets being poorly maintained or subject to unauthorized changes
- Controls not being implemented and operated
- Lack of control and discipline around the creation and governance of privileged accounts
- Lack of control and governance around the scheduling of batch jobs and backups, leading to unanticipated impacts for other providers and services
- Poorly thought-out metrics driving negative behaviors

The SIAM governance framework should also include controls to mitigate risks at the operational level.

2.3.5 Ownership of controls

Within a SIAM model, the customer organization always remains accountable and responsible for corporate governance, corporate risk management and governance of the service integrator. The customer organization should also be accountable and responsible for SIAM governance at the strategic level, including:

- Building and approving the SIAM business case
- Monitoring benefits realization (see section **2.2.5.2 Benefits realization management**)
- Approving the SIAM strategy
- Approving key design principles and aspects of the SIAM model, such as the capabilities it will source internally
- Ensuring compliance with regulations and other corporate governance requirements

For aspects such as designing the detailed SIAM model, process architecture and tooling strategy, the customer organization may choose to use external consultants and/or delegate the work to the service integrator. It should, however, be understood that this is only delegation of the responsibility for carrying out the design work, accountability for ensuring that a properly designed solution is produced and approved will remain with the customer organization.

Both the retained capabilities within the customer organization and the service integrator have an interest in ensuring that adequate controls are in place and that services are performing according to design. Controls allow responsibilities to be shared and, where appropriate, delegated to service providers. A success factor for this approach is ensuring that retained capabilities are provided with regular assurance that controls are in place, are measurable and are working. Ownership of controls needs to be agreed, assigned and understood, as well as being maintained over time as staff move roles or service providers change.

2.3.6 Governance framework

The controls and governance enablers should form a comprehensive governance framework, where the interconnection of the components allows them to support each other. **Figure 8** shows an example SIAM governance framework comprising different governance components and the relationships between them.

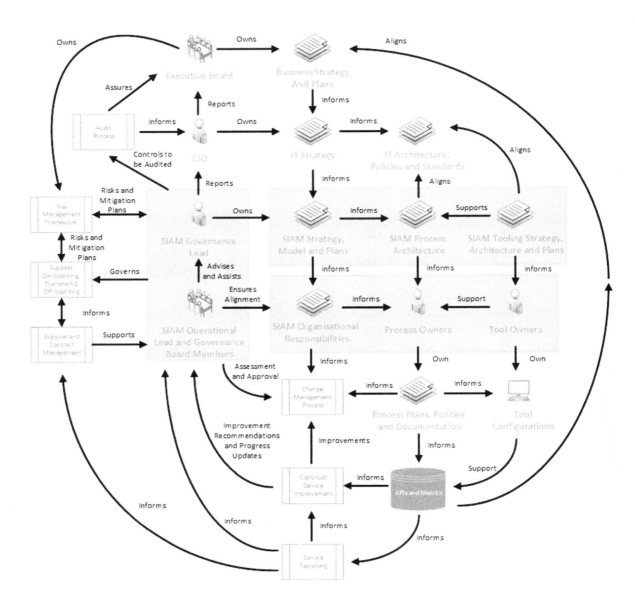

Figure 8: SIAM governance framework example[8]

The exact nature of the different governance components will vary depending on the organizations, their governance requirements and the SIAM model being adopted. There are many factors that need to be considered when designing an appropriate governance framework, including tradition, industry, size, maturity and culture. It is important to understand that attempting to force an overly bureaucratic and rigid governance framework onto a smaller organization with an informal culture is unlikely to be successful. Larger and more corporate organizations may expect more formal controls.

The governance framework, along with components within it, such as the SIAM strategy, should be reviewed and maintained regularly. This should consider changing business requirements, industry trends, new technologies and emerging threats. The framework or individual components may need to

---

[8] Governance in a multi-supplier environment, itSMF UK SIAM Special Interest Group/Neil Battell.

be updated or adapted to meet evolving requirements and the framework itself should provide the mechanism for doing this through established governance boards.

2.3.7 SIAM governance roles

It is important to understand the difference between the governance roles defined as part of the SIAM governance framework, and the delivery roles that perform activities within the SIAM ecosystem. The focus of the governance roles is to monitor, provide assurance and, when necessary, take decisions and/or actions to exert control and provide course corrections. The focus of delivery roles is to perform the day-to-day activities that are required for the ecosystem to function and for services to be delivered (see **Figure 23: Mapping SIAM roles onto the COBIT 5 business framework**).

In this section, the following areas will be discussed in further detail:

- SIAM Governance Lead
- SIAM Operational Lead
- Process Owner Roles
- Governance Boards

2.3.7.1 SIAM Governance Lead

The SIAM Governance Lead is a senior role residing within the customer's retained capabilities. They are primarily responsible for providing assurance regarding the implementation and operation of the SIAM strategy and operating model. The role requires the following knowledge, skills and experience:

- IT governance and risk management
- Knowledge or experience of auditing
- Experience of engaging service providers
- IT operations and large program management
- Service management
- Excellent communication and reporting skills
- Ability to communicate at all levels, across multiple organizations

The SIAM Governance Lead owns key SIAM governance artefacts, including the SIAM strategy and SIAM model, and is responsible for ensuring that these items are maintained in alignment with business requirements. Responsibilities of the role include:

- Ownership of strategic SIAM governance artefacts
- Setting up the SIAM governance board structure, assuring its successful and sustainable operation, and chairing governance board meetings
- Ensuring that all governance related roles are allocated, understood and performed
- Ensuring the design, definition and implementation of a SIAM governance framework appropriate to the customer organization, culture, the SIAM model being adopted and agreed risk profile
- Ensuring review and ongoing maintenance of the SIAM governance framework
- Working with the project board to oversee governance during transition to the SIAM model
- Working with the service integrator and service providers to ensure successful ongoing operation of the SIAM governance framework

- Identification and management of risks related to governance of the SIAM ecosystem
- Providing assurance to senior management that the SIAM ecosystem is being governed effectively
- Ensuring SIAM operations align to strategic objectives
- Ensuring regular measurement and review of both service integrator and overall performance
- Continual improvement of SIAM governance
- Providing guidance and leadership with regard to governance

The role should be fully defined and its responsibilities documented and agreed formally.

Delegated governance

Some customer organizations choose to have their retained capabilities responsible only for holding the service integrator to account and safeguarding the organizational objectives. The remaining duties (such as setting up governance boards, continual improvement, etc.) are delegated to the SIAM Operational Lead, a role often assigned to the service integrator.

2.3.7.2 SIAM Operational Lead

Responsibility needs to be allocated for leading and managing the overall operation of the SIAM ecosystem, providing direction and acting as the escalation point for any management issues. The SIAM Operational Lead role normally sits within the service integrator layer, although there may be some exceptions where the role resides within the retained organization because more involvement or control is desired.

SIAM Operational Lead

The role may often be performed by the person with a title such as 'Head of Service Delivery'.

The SIAM Operational Lead must work closely with delivery leads from different organizations within the ecosystem, along with the process owners and managers who are responsible for the processes being operated. The role will build, operate and improve service delivery across the ecosystem, ensuring that plans and objectives are communicated and understood, and issues resolved in a timely manner. In conjunction with the SIAM Governance Lead, the role is responsible for ensuring that operational governance is in place and followed.

2.3.7.3 Process owner roles

Process owners will exist within the customer organization's retained capabilities, the service integrator and the service providers. Within the service provider layer, they are responsible for governance of the process(es) within that organization, and alignment with the end-to-end processes operating within the SIAM model.

Within the service integrator, the process owner is accountable for the end-to-end governance and integration of the process or processes within their scope, ensuring that they are properly implemented, managed effectively, functioning as intended and working across all relevant providers and teams. The customer organization retains ownership of any processes that are not governed by the service integrator, for example, contract management.

Assigned process owners

There is not necessarily one individual owner per process. In practice, and depending on the size of the organization, one person could be the owner of several processes. These can be grouped by line of business, lifecycle stage, etc. Within the service integrator, these are sometimes referred to as 'service architects' rather than process owners.

The service integrator may need to provide support to resolve interface issues related to processes, for example where a 'downstream' service provider needs data from an 'upstream' provider. Providing this data may represent a cost with no immediate value to the upstream provider, so it may not be provided freely. The service integrator needs to be able to draw upon appropriate incentives to ensure that interface issues are addressed effectively and in a timely manner.

Process owners own their process documentation, ensuring that it is maintained and readily available to practitioners that are participating in the process. They need to ensure that all process interfaces are functioning and that, where necessary, improvement plans are developed and executed.

Process owners must work in conjunction with process managers to govern and manage processes, ensuring that:

- Staff are aware of the process, when to use it and how to trigger it
- Process documentation is well maintained and available to those who require it
- Process practitioners receive adequate and regular training
- The process, its policies and associated responsibilities are understood by all stakeholders
- Process roles are assigned and reviewed when required
- The process is monitored and controlled to ensure conformity
- Process outputs are of an appropriate quality
- Process risks are identified, assessed and managed
- Adequate resources and skills exist to allow the process to operate effectively
- All interfaces and integrations are functioning correctly

Being a process owner is not necessarily a full-time role. It is possible to be a process owner of multiple processes or combine the role with other responsibilities.

2.3.7.4 Governance boards

Within SIAM, boards are regarded as structural elements. These organizational entities have specific responsibilities and work across multiple organizations and layers in the SIAM ecosystem. Boards perform a key role in providing governance. They do this by acting as decision-making bodies, which are convened regularly throughout the operating lifespan of the SIAM model. Boards are accountable for the decisions that they make.

Governance

Corporate organizations often use the word 'governance' to describe both the way boards or their like direct a corporation, and the laws and customs (rules) applying to that direction.

For an organization embarking on the transition to a SIAM model, governance boards must be designed and implemented from the outset to ensure that an integrated, cohesive and clearly understood accountability structure is present for all service providers. Within the SIAM governance framework, there should be several different governance boards, each with its own remit in terms of areas of responsibility. The number and type of boards will vary depending on the size and complexity of the SIAM ecosystem, as well as the culture of the customer and service integrator organizations.

Board structures

A global organization may require a hierarchy of boards based on geographic areas, with different boards at country, regional and global levels. This decision will need to be based on the degree of variation in terms of services, infrastructures, service providers and cultures between the different geographies.

The board structure should provide governance at strategic, tactical and operational levels. Examples include:

- **Strategic:** approval of funding, contractual and commercial agreements and strategy
- **Tactical:** approval of policies
- **Operational:** approval of (minor) changes to services and processes (Note that a change to a service may appear operational, but depending on the scope and impact of the change it can be deemed tactical or even strategic)

The board structure should be designed to be mutually exclusive; collectively exhaustive (MECE) so that:

- **Mutually exclusive:** each likely issue or topic is within the purview of one, and only one, board
- **Collectively exhaustive**: the boards' terms of reference together address every aspect of the service to assure effective governance

Figure 9 shows an example of a governance board structure. Many of the boards are facilitated by senior service integrator roles (such as the SIAM Operational Lead Role), while other boards are run by staff in function-specific roles (such as the lead architect or change manager). SIAM governance retains oversight to ensure that the boards are in place and operating as intended.

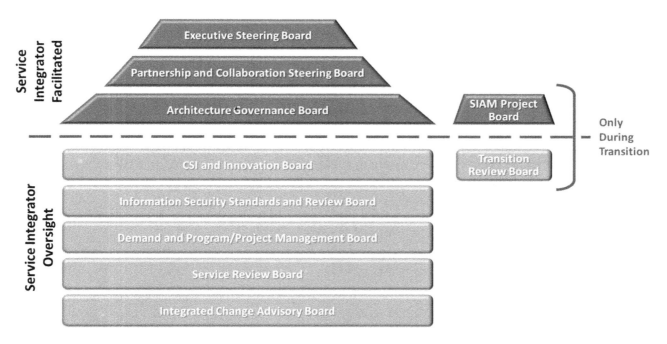

Figure 9: Illustrative types of governance boards[9]

Examples of different types of governance boards are described below.

Executive Steering Board

The senior board responsible for strategic decisions regarding the design, transition and operation of the SIAM model. The board sets and publishes policies that describe the governance of the SIAM organization structure. These must be made available to all parties and stored in a shared repository. See section **2.3.7.5** for more details on the **Executive Steering Board**.

Partnership and Collaboration Steering Board

A joint board, focused on building partnerships and collaboration between the different parties within the ecosystem. The board seeks to promote, encourage, track and reward desired behaviors.

Architecture Governance Board

A board with representation from across the ecosystem, responsible for creating, agreeing and maintaining architectural standards that must be adhered to across the ecosystem. It will also review proposed service designs, identify necessary changes and approve final designs and exception waivers.

This board can be formed from the architecture function of the transition project team, as discussed earlier in this publication. The board is also responsible for actively seeking to encourage technology innovation, understanding new technologies and reducing technical debt.

---

[9] Source courtesy of Chris Taylor-Cutter.

CSI and Innovation Board

This is a joint board responsible for promoting and tracking continual service improvement and innovation, allowing service improvements to be investigated, prioritized, planned and implemented.

Information Security Standards and Review Board

A board responsible for agreeing information security standards in line with the customer organization's security policies to be applied across the SIAM ecosystem. It will review proposed designs for new or changed services.

In many organizations, this board would exist outside the SIAM ecosystem. It would set policies and standards to be implemented by the operational boards.

Demand Board

A board responsible for reviewing and aggregating demand, understanding and agreeing priorities, and managing the allocation of resources to programs and projects. Depending on the position of the SIAM ecosystem within the IT function, this board may be positioned at a higher level, outside of the SIAM governance framework.

Service Review Board

A board responsible for jointly reviewing service level achievement, approving short- and long-term plans and activities regarding cross-provider service delivery, resolving cross-service provider delivery issues, identifying and, where necessary, escalating opportunities and issues, and managing cross-service provider risks (see section **5.1.2 Tactical governance boards**).

Integrated Change Advisory Board

A joint board responsible for reviewing, assessing and advising on proposed changes that may have an impact across multiple providers, are high risk or 'significant' in terms of size or complexity.

In addition to the Change Advisory Board, there will be process-specific working groups and forums acting at an operational level. The groups and forums build cross-service provider partnership and collaboration between SMEs working in process areas (see section **1.7.4 SIAM structural elements**).

Board membership must be defined, allocated and agreed, including chairs and vice chairs. The frequency of meetings, schedules, logistics (locations, virtual or face to face, circulation of material for discussion in advance, etc.) and standard agendas must be defined. Terms of reference must be developed for each board, defining its remit, scope and level of authority. Meeting etiquette descriptions can help to set expectations and should be included as part of the process for onboarding new staff and service providers (see also section **2.3.13.4 Onboarding and offboarding of service providers**).

The service integrator must agree the SIAM governance framework (including the board structure) during the design of the SIAM model, in collaboration with the customer organization. The framework often changes during implementation. It is important to set out the desired target model and to be prepared to be flexible when required.

2.3.7.5 Executive Steering Board

Given the important role played by the Executive Steering Board, it is worth exploring its purpose in more depth. Its remit is to:

- Ensure strategic and operational business alignment between the parties in the SIAM model
- Analyze customer, service integrator and service provider business plans, and oversee new or modified services
- Develop strategic requirements and plans associated with services
- Conduct periodic review of governance boards and members
- Resolve issues and exceptions escalated by other governance boards
- Review and agree the assessment of the service integrator's performance and that of the service providers
- Provide oversight (and assurance) of:
  - Transition plan, progress and achievement of critical deliverables and key activities
  - Operational performance against service levels and continual improvement adjustments
  - Governance, risk and compliance obligations against policies, standards and regulations
  - Security reporting, vulnerability and compliance dashboard
  - Financial performance and forecast
  - Effective risk management and monitoring, including remediation of audit actions
  - Tracking against contracted deliverables and obligations

The following is an example agenda for an Executive Steering Board meeting:

- Minutes and actions of last meeting
- Review authority and membership of other governance boards
- Discuss all escalated issues
- Discuss and agree business goals and objectives, priorities and strategy
- Review reports on performance, customer satisfaction and financial issues
- Review results of internal and/or external audits
- Review the risks and issues log
- Celebrate success (many of these meetings are perceived to be 'dry', so it can be good to add some positivity and show that there are many things working well)
- Any other business

The Executive Steering Board must ensure that each service provider adheres to relevant company laws and regulation in the delivery of services to the customer organization, as well as to its obligations under contract. It must also confirm that business decisions affecting the SIAM ecosystem have been authorized correctly.

Company law

Each jurisdiction has specific requirements for company operation and management of people, along with governance and reporting. Where the scope of the SIAM implementation crosses jurisdictional boundaries, all applicable legal considerations must be considered.

For example, for organizations based in the UK, this may entail the Executive Board gaining assurance that all service providers are incorporated and administered in accordance with the Companies Act and general law.

Another example is that companies that are registered in the USA are subject to Sarbanes-Oxley (SOX) board-level reporting.

Compliance with required legislation is the responsibility of each service provider. The Executive Steering Board must be provided with confirmation that the service providers in each layer of the SIAM ecosystem adhere to all relevant legal requirements. Periodically, there may be changes to national or international law that affect the services and the obligations of the parties. Compliance will need to be maintained by all parties, across all applicable laws and any changes.

Changes in law

The introduction of the General Data Protection Regulation (GDPR) within Europe affects any organization trading with a European customer.

The regulation was adopted on 27 April 2016 and became enforceable on 25 May 2018 after a two-year transition period. Unlike a directive, it does not require national governments to pass any enabling legislation and is thus directly binding and applicable.

The board should have membership from senior executives within the customer organization as well as representation from the service integrator and each of the core service providers. Core service providers are those providers that are taking an active role within the SIAM ecosystem. There may be other service providers (for example, those simply providing commodity-as-a-service solutions) performing a passive role. These would not be expected to regularly attend governance boards (nor would they be likely to be interested in doing so).

Care must be taken to ensure boards are an appropriate size. Too many attendees can make a board unwieldy, whereas too few will lack legitimacy. Members need to have sufficient authority to provide effective oversight of the outputs of the SIAM model. Attendees are expected to:

- Represent and explain their organization's performance and future plans
- Assess risks and issues to service, customer experience, regulatory concerns and other considerations, and contribute to their mitigation and resolution
- Discuss commercial and/or sensitive issues to assure the quality of the end-to-end SIAM ecosystem

The Executive Steering Board focus areas are:

- Policies
- Governance processes
- Ways of working
- Regulatory considerations
- Management and control methods

Policies

The Executive Steering Board is accountable for ensuring that policies relating to the delivery of services within the SIAM ecosystem are produced, and are accurate and maintained. The board may not directly create these policies, but it does ensure that appropriate policies exist and are followed.

The policies must be agreed by all service providers, with escalation to the Executive Steering Board if conflicts arise or changes are identified.

The policies must conform to the organization's quality management standards and procedures, and be subject to review at regular intervals.

Example governance areas requiring policies include:

- Financial governance:
  o Application of financial controls to the SIAM ecosystem and its service providers
  o Commercial aspects of contracts and post-contract delivery
  o Timely credit control and cash management policies
- Governance risk and assurance:
  o An effective culture of risk awareness and management maintained throughout the SIAM environment
  o Service commitments are independently reviewed and approved, with respect to profitability and risk
  o Regular monitoring of effective management of delivery
  o Adherence to approved business processes through audit
  o Continual improvement of business processes
- Corporate services governance:
  o Conformity to relevant laws and ethics
  o Business decisions supported by relevant legal advice
  o Effective and efficient provision of facilities
  o Conformity to procurement policies and processes
  o Compliance with contractual commitments

Processes

The Executive Steering Board is also accountable for ensuring that the processes by which the SIAM model will be governed are defined, implemented and controlled. Performance and compliance with the agreed processes must be reviewed by relevant working groups and forums, with issues escalated to the Executive Steering Board for evaluation and guidance. Where process failures have been identified, corrective action and, potentially, service improvement activities, should be planned and implemented.

Ways of working

Often known as 'culture' or 'custom and practice', ways of working pertains to the methods by which services are delivered to the customer organization.

Attention must be paid to collaboration requirements from each service provider, and the Executive Steering Board must participate fully in setting the vision for the end-to-end service and the behaviors that must be displayed by each stakeholder.

Examples of planning, instilling, managing and evaluating ways of working include:

- Implementing relationship management processes – particularly focusing on interfaces between service providers and the 'handoff' procedures that must exist to protect end-to-end service delivery. This is in addition to processes such as contract management, supplier management and performance management.
- Agreeing methods by which people are transferred between service providers, whether internal or external. Attention must be paid to changes in roles and objectives, the commercial basis of delivering the services and the drivers that may affect decision-making processes (see also section **3.3.3.4 Outgoing resources**).
- Understanding that there can be resistance to change from some stakeholders. This resistance is often based on uncertainty, but is nevertheless a legitimate concern that needs to be addressed.
- Recognizing the benefits of taking time to communicate to all affected people, enabling them to understand the objectives, outcomes and expected behavior to support the SIAM model.
- Evaluating and publicizing changes to scope. The SIAM model will change the responsibilities and accountabilities of both organizations and people – there is usually an increase in the level of complexity, and so the need for good definition and clarity around who does what is essential.[10]
- Applying collaboration agreements, which define a basis from which service providers work together to deliver the end-to-end service (see section **3.1.8 Collaboration model**).

Regulatory considerations

Depending on the customer organization and its sector, attention must be paid to the external standards and regulatory requirements that affect services delivered by the SIAM model.

Financial Conduct Authority

As an example, for organizations operating in the financial services sector in the UK, the governance model must conform to the Financial Conduct Authority's (FCA) requirements, as described here:

"FCA regulations require that financial services (FS) businesses maintain effective control over their operations and closely manage all aspects of risk. The decision to use cloud services (or any third-party service provider) is just another risk that FS businesses must assess, quantify, justify and manage from the outset and as service provider relationships develop."[11]

The FCA has recently issued guidance for companies outsourcing to the cloud and other IT services (FG 16/5), expanding on its existing outsourcing rules.

---

[10] Sarbanes-Oxley, 2016.

[11] For more information, visit: *www.handbook.fca.org.uk/handbook*.

So, what do you need to think about?

- **Control and data security**: These are at the forefront of the FCA's concerns. The outsourced process is still your responsibility and companies need to retain sufficient expertise to manage the outsourcing.
- **Diligence**: Is there a good business case for using the cloud? Is the provider reliable and competent? How will you monitor this? What will you do if it all goes horribly wrong?
- **Transition and contingency:** Can you effect a smooth transition to another provider, and properly manage and minimize the effect of outages?

It is not always easy to apply existing rules in a new and dynamic context, and the guidance has led to some interesting practical questions, such as:

- **Jurisdiction**: Can you control where your data is processed? The FCA suggests that businesses agree a data residency policy.
- **Access to premises**: The FCA still requires that companies, auditors and the Regulator should have physical access to service provider's premises to monitor performance. This poses challenges with highly secure cloud facilities, but it is up to the customer organization to determine with the cloud provider how this access can be achieved.

Management and control methods

The Executive Steering Board is accountable for the definition and management of a consistent, shared vision for the SIAM ecosystem. This means that the SIAM operating model considers the impact across all stakeholders, embodied by:

- Consistent and well understood cross-service provider processes
- Contracts and service levels with each service provider that support the SIAM model, and cooperative working practices across service providers
- Effective governance boards, processes and controls that enable the customer organization to manage risk, exercise appropriate control and provide direction
- A business change lifecycle that ensures early consideration of how a new service integrates with existing services
- A consistent understanding of the roles of the stakeholders through communication and education
- The empowerment of the service integrator by the customer, with a zero-tolerance policy to non-adherence to the model and processes, ensuring that service providers interface with and support the service integrator appropriately

2.3.8 Ownership

Many issues can be traced to a lack of clear ownership. Over time, toolsets, documentation, targets, monitoring solutions, controls and reports may become out of date because of a lack of clear ownership and responsibility for their upkeep. Ownership is a key governance concept within a SIAM ecosystem. The number of service providers and changes to service providers creates a volatile environment. A lack of clarity may lead to increased risk because of incorrect assumptions being made.

A grey area

While conducting an assessment within a SIAM model, consultants found that the term 'that's a bit of a grey area' had become part of the organization's common corporate language when asked about ownership and responsibilities. The expression was heard in almost every interview.

Artefacts within the SIAM governance framework, such as strategies, plans, policies, toolsets and processes, all need to have clear and appropriate ownership assigned, agreed and documented. Ownership needs to be assigned to appropriate roles as it is active and ongoing. There is little value in assigning ownership at either too junior a level (with no experience or authority) or too high a level (with no visibility or focus on the component being owned). The role with the responsibility for owning and maintaining a governance artefact also needs to have access to sufficient budget to support its activities.

2.3.9 Segregation of duties

Segregation of duties is an important governance concept that needs to be considered when defining the approaches to be used within the SIAM model and associated processes. Appropriate segregation of duties in processes, procedures and tasks ensures that two or more people are always involved in the end-to-end completion of the activity. This is particularly important for activities that might be subject to fraud or other tampering. Segregation of duties provides oversight and helps ensure that errors are identified, while also making fraudulent activities significantly more difficult to conduct.

Segregation of duties

For an internal service integrator SIAM structure, there should be segregation of duties between the internal service integrator role and the retained capabilities. This avoids potential issues, for example, the perception that internal providers are treated preferentially.

This also applies to the Lead Supplier structure, when the role of the service integrator and the service provider needs to be segregated.

Examples of common segregation of duty include:

- Orders raised by one person need to be signed off by a different person
- Requests for new or replacement equipment need to be approved by someone other than the requestor
- Changes to critical services need to be peer reviewed before being released
- Change raisers are not permitted to approve their own changes
- Travel expenses need to be approved by line managers
- Staff in an internal service integrator report to a different management line from the staff in internal service provider roles, to avoid the perception of conflict of interest
- Privileged information available to the service integrator is not shared with an internal service provider unless it is also shared with external service providers
- Reporting on an internal service integrator's performance is kept separate from reporting on internal service providers

Conflict of interest

In an early implementation of SIAM in a very large organization, there was significant concern about the perception of conflict of interest between the proposed external service integrator and the service providers. There was even a suggestion that the people who bid for the service integrator contract should be excluded from bidding for service provider contracts.

A conflict of interest (CoI) management plan was proposed to allow a single organization to bid for multiple contracts. This included specific training on the CoI plan for all service integrator staff and the path to escalate inappropriate requests and approaches.

Often, controls such as segregation of duties are standard managerial controls that extend beyond a SIAM model and already exist. They need to be reviewed, adopted and included within the SIAM model (see section **2.4 Define principles and policies for roles and responsibilities** and **3.1.6 Detailed roles and responsibilities**).

2.3.10 Documentation

Another governance consideration is how to store, control and maintain key SIAM-related artefacts, such as the SIAM strategy, SIAM model definition, plans, process definitions, contracts and policies. These documents contain valuable information that needs to be readily available to the appropriate personnel, without them having to waste time searching for it.

There is also a need to prevent any unauthorized changes to the documents and to restrict access based on roles and responsibilities.

Document classification

In one example, the service integrator created a schema of classification of data and associated access rights. All operational data types were assigned a default classification. This greatly reduced the effort associated with managing access to documents and formed the basis of access policies.

2.3.10.1 Document storage and access control

Different organizations have different approaches to document storage, ranging from unstructured shared drives to complete document management systems with built-in version control and workflows.

Document availability

Controlled documents are often held on corporate intranets, which may not be accessible by outside organizations, for example, external service providers. This can lead to multiple copies being downloaded and stored, making it impossible to know which version anyone is using at any one time.

SIAM project documentation governance requirements need to be identified. Any existing document management solutions can be assessed to see if they are fit for purpose, or a new solution may be put in place.

2.3.10.2 Document reviews and change approvals

Documentation within the SIAM governance framework (such as process and policy definitions) must be reviewed periodically, at least annually and when affected by changes. Reviews ensure it is still aligned with business requirements and remains current. Where necessary, changes are made, and the new document version approved and stored appropriately. It is the responsibility of the respective document owner to ensure that reviews occur.

Although some organizations place important documentation under the control of formal change management processes, this is still not common practice. Serious consideration should be given to using any formal, existing or planned change management process, as this is likely to have impact assessment, approval and notifications of change within its workflow, as well as supporting tooling. Alternatively, organizations may turn to other solutions, such as document management tools, or in smaller organizations, simple procedures involving new document versions being emailed to members of the respective governance board for review and approval.

2.3.10.3 Documentation requirements

Governance of SIAM documentation should ensure that:

- Documents have been created, approved formally and signed off by the appropriate authority
- Documents have an assigned and active owner
- There is assurance that the version being used is the latest definitive issue
- Documents are reviewed periodically for alignment and currency
- Documents are easy to locate and visible to authorized users
- Documents cannot be viewed by unauthorized users
- Unauthorized changes cannot be made to the documents
- Changes to the documents are tracked formally, reviewed and approved

2.3.11 Risk management

The SIAM governance framework should include policies, procedures and responsibilities for risk management. Risk management is the process of identifying and assessing risk, and taking steps to reduce risk to an acceptable level, where appropriate. It is common practice to use the customer organization's risk management approach as a mechanism to support the transition to a SIAM model.

The organization may already have an approach for enterprise risk management. Even if that is the case, SIAM governance should ensure that there is an appropriate and effective mechanism for identifying, understanding, tracking and managing specific risks to the SIAM ecosystem. Once this is established, information regarding the level of risk may then be shared with enterprise risk management, if required. If the customer organization does not have an established approach, there are best-practice techniques available, such as M_o_R® or the ISO 31000 standard.

Management of risk

ISO 31000:2018 Risk management – Guidelines provides principles, framework and a process for managing risk. It can be used by any organization regardless of its size, activity or sector. Using ISO 31000 can help organizations increase the likelihood of achieving objectives by improving the identification of opportunities and threats, and effectively allocating and using resources for risk treatment.[12]

The chosen approach determines the processes, techniques, tools and roles and responsibilities for risk management. The risk management plan describes how risk management will be structured and performed on the project. Governance needs to ensure that:

- Risk management policies, procedures and a matrix, or an agreed method to prioritize risks, are defined, documented and approved
- Roles and responsibilities regarding risk management are clearly defined and allocated
- Compliance with defined risk management policies and procedures is made part of service provider contracts
- Time is allocated during process and procedural design, planning sessions, team meetings and reviews to consider, identify and discuss potential risks
- Identified risks are recorded, categorized, tracked, owned and analyzed
- Significant risks, above the organization's risk appetite, are addressed, even if this is 'merely' a recorded decision by management to formally accept the risk
- Contingency plans are made where necessary for the occurrence of particular risk events
- Risks are reviewed regularly in case their nature, likelihood or potential impact has changed

The first step in creating an effective risk management system is to understand the distinctions between the types of risk that organizations face, for example:

1. **Preventable risks**

    These are internal risks, arising from within each organization, that are controllable and ought to be eliminated or avoided. Examples are risks from employees' and managers' unauthorized, illegal, unethical, incorrect or inappropriate actions, and the risks from breakdowns in routine operational processes.

    Each organization must define 'a zone of tolerance' for defects or errors that would not cause severe damage to the enterprise or transition project, and for which achieving complete avoidance would be too costly. In general, they should seek to eliminate these risks.

    This risk category is best managed through active prevention, monitoring operational processes and guiding people's behaviors and decisions toward desired norms. There are many sources of information available to guide an organization through such a rule-based compliance approach, which provide adequate guidance for use within a SIAM project.

---

[12] *www.iso.org/standard/65694.html*.

2. Strategic risks

Strategic risks are quite different from preventable risks because they are not inherently undesirable. The customer organization will voluntarily accept some risk in order to generate superior returns from its strategy.

A strategy with expected high returns generally requires the organization to take on significant risks, and managing those risks is a key driver in capturing potential gains. For example, BP accepted the high risks of drilling several miles below the surface of the Gulf of Mexico because of the high value of the oil and gas it hoped to extract.

Strategic risks cannot be managed through a rule-based control model. Instead, these risks require a risk management system designed to reduce the probability that the assumed risks materialize and to improve the organization's ability to manage or contain the risks should they occur. Such a system would not stop organizations from undertaking risky ventures, to the contrary, it would enable them to take on higher risk, higher reward ventures than they could with less effective risk management.

A transition to a SIAM model can be considered as a strategic risk, for customers, service providers and the service integrator.

3. External risks

Some risks arise from events external to the company and are beyond its influence or control. Sources of these risks include natural and political disasters and major macroeconomic shifts. As a transition to a SIAM model usually increases the number of different parties in an ecosystem, the external risks are likely to increase as well.

External risks require a different approach. As companies cannot prevent such events from occurring, they must focus on proactive identification and mitigation of any impact. Consider examples from the organization's history and scenarios of its possible future, which will enhance communication and help executives understand more easily the context in which they are operating.

The scope of SIAM

Techniques such as the Cynefin™[13] approach are useful here. Cynefin allows decision makers to see things from new viewpoints, assimilate complex concepts and address real-world problems and opportunities.

Organizations should tailor their risk management processes to these categories. Although a compliance-based approach is effective for managing preventable risks, it is wholly inadequate for strategic risks or external risks, which require a different approach based on open and explicit risk-based discussions.

A control needs to be defined for each risk. Viable, effective controls will be:

---

[13] *http://cognitive-edge.com*.

- Appropriate to the risks they are addressing
- Aligned to the strategy
- Cost justifiable
- Stable and repeatable, so that they operate in the expected manner each time
- Verifiable, so that it is possible to check the existence and operation of the control
- Collaborative, so that key stakeholders are part of decision making, control design and control implementation
- Owned, so that someone is responsible for the control being implemented and operated as designed

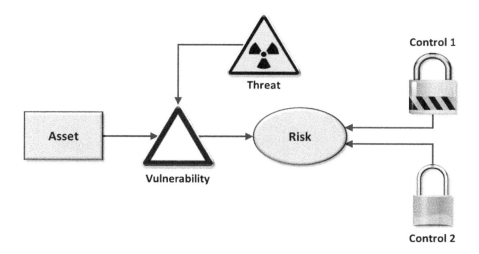

Figure 10: Assets, threats, vulnerabilities, risks and controls

Figure 10 shows the relationships between assets, threats, vulnerabilities, risks and controls. A risk occurs when there is an asset that has a vulnerability to a particular threat. Where the size of the risk is above the organization's risk appetite, and the asset or threat cannot be removed to eliminate the risk, one or more controls should be put in place to mitigate the risk.

It is good practice to use a combination of different control types when addressing a risk. Common control types are described in **Table 1**:

Table 1: Common Risk Control Types

Control type	Description	Examples
Directive	Controls that reduce risk by providing direction and guidance	• Policies • Procedures • Standards
Preventative	Controls that reduce risk by preventing or at least reducing the likelihood or impact of the risk occurring	• Automation • Training • Approval processes • Passwords • Segregation of duties
Detective	Controls that detect when a risk event is either occurring or about to occur	• Monitoring expenditure against budget • Event monitoring toolsets • Trend analysis on metrics • Audits/sampling
Corrective	Controls that trigger some form of corrective or recovery action should a risk event occur or be about to occur	• Automated fail-overs • Defined procedures followed for security breaches • Recovery plans

A number of controls may be aimed at managing the risk of 'expected benefits not being realized from the SIAM program'. Example controls are illustrated in **Table 2**:

Table 2: Control Example Managing the Risk of 'Expected Benefits Not Being Realized from the SIAM Program'

Control	Control type	Purpose
Formal review and approval of business case and expected benefits	**Preventative**	To help ensure that the business case and its expected benefits are appropriately defined, of sufficient quality and are achievable
Program change approval process	**Preventative**	To prevent changes being made to the objectives, scope or execution of the program without proper consideration of the impact on expected benefits
Regular monitoring of expected and realized benefits	**Detective**	To determine whether benefits realization is on track to achieve expected results

Control	Control type	Purpose
both during and after the program		
Regular benefits realization reviews	**Corrective**	Regular reviews of expected and realized benefits to identify whether any corrective actions or initiatives need to be undertaken to help ensure targets are achieved

The risk management processes for the SIAM transition project will be similar to those that are to be used within the operational SIAM model, and can be handed over as part of the Implement stage (see section **4.2 How to transition to the approved SIAM model**).

2.3.12 Auditing controls

The SIAM Governance Lead should work with internal audit personnel to ensure that key controls are audited regularly to provide assurance that required controls are in place and functioning effectively. Relevant audit policies, plans and schedules should be developed and maintained. Any audit issues identified must be owned and monitored through to resolution.

Some organizations may choose to adopt standards such as ISO/IEC 20000, either purely as a reference or with the aim of achieving certification against the standard. When adopting a standard, it is important to check whether it has been written in a way that allows it to be used within a SIAM ecosystem.

2.3.13 Supplier and contract management

The management of suppliers (including external service providers) and contracts needs to be a core capability within any SIAM ecosystem. The aim is to build an overall culture of partnership, collaboration and innovation rather than one of 'rule by contract'. 'Rule by contract' is a situation where frequent debates occur around what is and is not agreed in contracts and the additional costs that may be charged or the penalties that may be levied. Ideally, better supplier and contract management starts from a foundation of cost transparency and fair pricing.

An example of unfair pricing

There is little point in entering into a contract with a service provider where prices have been driven down to such a level that the service provider cannot possibly even break-even – unless they charge extra for every small additional request not covered in the contract.

This approach almost always leads to a poor relationship, where the customer organization feels that everything it asks for comes with a price, and the service provider becomes frustrated because any small innovations or improvements it suggests are not progressed because they would have to be charged for, since the contract margins are so low.

Often, one of the aims of adopting SIAM is to move from a situation where the customer organization is very constricted, possibly by a single, monolithic outsourcing contract, to a situation where it gains more flexibility. This allows it to place different services with whichever service provider – internal or external – has the best offering or most attractive price point at the time. This flexibility requires due consideration within the strategic and design activities where EA considerations are planned and created. The ability to easily onboard, transition between and offboard service providers as requirements or offerings change, is one of the key benefits of a SIAM model.

SIAM contracts often have a shorter duration to prevent lengthy tie-ins, and are structured to allow flexibility in terms of future changes to services and how they are delivered. However, in some instances, the nature of the relationship required between the customer and service provider requires longer term contracts to provide enhanced relationships and a greater degree of stability.

2.3.13.1 Supplier and contract management function

The supplier and contract management function needs to subscribe to and support the SIAM model. Therefore, it is important that cooperation is established as early as possible in the SIAM implementation, between the project team and the supplier and contract management function. The project team and, once appointed, the service integrator, should obtain input and guidance from the supplier and contract management function throughout the entire lifecycle of a service provider contract. Typically, the service integrator will have a central role in aspects such as change management, service performance management and supplier management, working with the customer-retained capabilities' contract management function.

Often, supplier and contract management will belong to a part of the organization outside IT, but this activity cannot be completely isolated from IT. Adopting a SIAM model is an organizational and cultural change initiative. SIAM often has a role in integrating services that extend beyond IT but rely upon IT, such as finance and accounting, human resources (HR) services or other business processes. As such, the whole organization, including senior management and the supplier and contract management function, needs to understand the benefits and be fully committed to the program.

2.3.13.2 Supplier and contract restrictions

Organizations planning their SIAM operating model and the future service provider landscape may want to establish design principles. These can be converted into constraints, which may restrict the scope of service that an individual service provider can bid for or take on. This prevents unhealthy monopolization by one service provider, either internal or external. Constraints may also prevent a service provider bidding across layers, for example, being the service integrator and a service provider (so removing the possibility of a Lead Supplier structure).

Although it is not uncommon in SIAM environments for organizations to be both a service integrator and a service provider, some customer organizations may make the decision to prohibit this. This allows the service integrator to focus on its integration responsibilities with no question of unfairness or potential conflicts of interest between the service providers (see section **1.6.4 Lead supplier as service integrator**). Note that it can sometimes be commercially difficult to achieve profitability as a lone service integrator, and therefore this function is often bundled with other services, albeit with separate contracts.

2.3.13.3 Supplier and contract governance

The governance of contracts with key service providers needs to involve regular review meetings and measures to assess personnel changes. This ensures that when changes occur, in either the customer organization or a service provider, new people are engaged as early as possible and brought into the preferred way of working to assure consistency of approach (see also section **3.1.2 Sourcing approach and the selected SIAM structure**).

Governance needs to ensure that:

- There is a defined and effective framework for managing service providers and contracts that is aligned with the specific requirements of SIAM
- There are appropriate levels of contract management resources, knowledge and skills in place to enable the framework to be designed and operated successfully
- Roles and responsibilities are clearly defined and allocated
- Contracts consider the full contract lifecycle, including responsibilities and deliverables
- Governance of key service providers and contracts is active and effective
- Contracts identify the service integrator as the agent of the customer organization, whether it is internally or externally sourced
- Service provider performance is measured and reported at agreed intervals against contracted commitments and service performance measures, with an agreed approach to address any shortfalls
- All service providers within the ecosystem are treated equitably
- Service provider targets are aligned with end-to-end service levels
- Controls are in place to ensure that contract review, renewal and end dates are tracked, so that timely action can be taken where necessary
- Steps are taken to prevent contractual relationships being affected when personnel or senior stakeholders change in any SIAM layer
- Contracts, wherever possible, follow a standard template with well-defined terms and conditions, escalation and dispute procedures
- Contracts incorporate clauses that promote collaboration and innovation
- Contracts allow for future flexibility in terms of changes to services and delivery mechanisms

2.3.13.4 Onboarding and offboarding of service providers

As part of supplier and contract management, significant attention and planning must be given as to how:

- New service providers will be onboarded into the ecosystem
- Service workloads will be migrated from one service provider to another at the end of the contract
- Outgoing service providers will be offboarded

The effectiveness and efficiency of these transition projects is a good indicator of the maturity of the SIAM operation. These transition points involve risk, so the SIAM governance framework needs to ensure that they are identified, understood and managed. There should be defined procedures for onboarding and offboarding service providers, and for managing the transition of service workloads between different service providers. The procedures should be supported by clauses in service provider contracts.

Each time a service provider is onboarded or offboarded, it should be treated as an improvement opportunity with the corresponding procedures being improved over time, as lessons are learnt.

2.3.14 Monitoring and measuring service performance

An important part of SIAM governance is being able to monitor, measure and understand how the ecosystem is performing. Without pertinent, accurate, up-to-date information, it is impossible to make timely and correct decisions, trigger effective corrective actions and understand whether plans are being successful.

Service performance measures whether a service is meeting defined business requirements. Trend analysis is vital as it is difficult to judge performance by individual numbers or metrics. Whether performance is improving also needs to be assessed.

With a SIAM environment, there should be a shift from satisfying contractual targets to focusing on end-to-end service performance, innovation, collaboration and meeting the customer organization's changing needs. That is not to say that measurement against contractual service targets held in SLAs or operational level agreements (OLAs) is not important, but it is a basic expectation, not the primary focus. It is still important to understand and govern performance against service targets. This may include a whole range of service considerations, such as availability, incident response and fix times, resolution times for common requests and actual system response times. These measures contribute to demonstrating that business needs are being met, and may indicate opportunities for sharing innovation, good practice and lessons to be learnt.

In a SIAM ecosystem, measuring performance is important to both the customer-retained capabilities and the service integrator, but for different reasons. The customer organization is typically interested in:

- Overall performance of the SIAM model and the services being delivered, to understand whether the strategy is working successfully, providing value and meeting its needs.
- Performance of the service integrator, to understand if the service integrator is performing as expected and achieving the levels of collaboration, integration and innovation it is required to deliver.

The service integrator is responsible for providing assurance of the performance of individual service providers and the end-to-end service, so that expected outcomes are delivered to the customer organization.

The service integrator will be interested in:

- The overall status of each service being delivered, to understand how the service is performing when aggregated across all relevant service providers.
- Individual service provider performance against service targets, to understand where issues or opportunities for improvement may exist. For example, if one service provider is consistently attaining better results in a particular area, this may indicate a 'good practice' that can be discussed and potentially adopted by other service providers.

Although it is commonplace for contracts to have clauses that enable remedies to be applied for the failure to meet agreed service targets, it is important to remember that a SIAM ecosystem is built on a culture of collaboration and partnership, rather than a culture where there is regular debate on whether penalties should or should not apply.

When service targets are not met, the initial focus should be on working together to drive improvement and resolve any issues. Contractual remedies should be applied consistently, to ensure that all service providers are treated equally and without any ambiguity. It is desirable to reward service providers that demonstrate a willingness and capability to improve.

Penalties related to lost value

Some contracts attempt to make the remedy relative to the service failure (for example, the loss of value to the customer). If there is no relation between the failure and the loss, it can cause the service provider to resent a 'massive' penalty for a 'minimal' failure. A remedy deemed to be disproportionate could result in the reduced impact of the remedy, in terms of loss recovery and its ability to act as a deterrent.

However, this approach can also lead to issues, as it confuses penalties for service failures with consequential loss or damage. Consequential loss or damage is defined by specific legal requirements usually covered separately in a contract.

Service targets need to be created carefully, because metrics drive behavior – also known as the 'observer effect'. People do what they think they need (or want) to do, and are more likely to pay greater attention if they know it may be inspected. However, rigid interpretations of some targets can lead to counterproductive behaviors, especially as pressure to meet the target increases.

People do what you inspect, not what you expect[14]

A service desk had a metric to answer 100% of incoming phone calls within a maximum of three rings.

The service desk staff soon learned not to pick up any phone that rang more than three times, leaving it to ring until the caller hung up, allowing them to achieve their target.

When designing a new measure or target, consider both the positive and negative behaviors that might arise as a result. Service providers need to adhere to the '*spirit*' of the agreement as much as the '*letter*', and targets should be defined in such a way as to encourage this.

---

[14] *Who Says Elephants Can't Dance?* by Louis V. Gerstner Jr., Harper Collins, 2009.

Problem management targets versus outcomes

An example might be setting a target of reducing the number of open-problem records for a particular service, month on month.

Although the target might be set with good intentions, expecting to drive service improvement, it might result in the teams responsible for the service deciding not to log new problems when they find them, as that will reflect negatively on the metric.

Governance needs to ensure that:

- Defined policies and procedures are in place to monitor, measure and report on performance.
- Roles and responsibilities are clearly defined and allocated correctly.
- Appropriate toolsets are in place and configured to support monitoring, measurement and reporting of performance.
- Adequate resources are allocated.
- Performance metrics and reports are reviewed regularly and, where necessary, action is taken.
- All targets set are SMART (Specific, Measurable, Agreed, Relevant and Time bound).
- Metrics, measures and reports are maintained and updated or changed when necessary to ensure they remain relevant to business requirements.

As collaboration and partnership between the different organizations is one of the differentiators of a SIAM environment, it follows that this should be an important measure of how an organization is performing within the ecosystem. The challenge is defining what 'collaboration' means and how to measure it. This is not an activity that is easy to define, and different organizations will have different ideas regarding how to approach this.

A few examples are:

- Measuring attendance to collaborative process forums, for example, if a major service provider is regularly attending and participating in the governance boards
- Conducting and assessing periodic collaboration questionnaires
- Asking each organization to nominate the service provider within the ecosystem that they feel has been most collaborative during a measurement period, and recognizing the organization with the most nominations
- Monitoring the number of disputes within the last quarter related to each service provider
- Using measures to build a balanced scorecard approach, perhaps including:
  - Number of service improvements implemented
  - Customer satisfaction scores improving
  - Changes in volumes, for example, a reduction in incidents, an increase in problems being identified or a reduction in failed changes
  - Innovation activities

Another difficult consideration can be how to measure and compare the performance of service providers when they are delivering different services. One service provider may receive more incidents than others, but, if it is responsible for an archaic legacy service or a service that is undergoing a significant level of change, it may be justifiable. Another service provider may always take longer to fulfil requests, but this could be because of the nature of the requests it receives.

The measures and targets set must always be appropriate to the individual service provider and the services it is supporting. Then, service providers can be compared in terms of how often each provider has achieved or missed its own targets.

2.3.15 Demand management

The SIAM governance framework must include controls to ensure that demand is recognized, validated, aggregated, monitored and allocated appropriately. Without these controls, there will always be a tendency for:

- Business units requiring new services to approach the service providers they are most familiar with, rather than those that may be able to offer the most cost-effective delivery.
- Requests not being aggregated, leading to a complex spread of many similar but non-standard services, rather than standardized services that take advantage of economies of scale.

Governance needs to ensure that:

- Demand requests for new or changed services are fed into a central point where they can be analyzed and understood (for example, the service integrator or the customer organization's retained capabilities).
- Current and future demand for services is monitored, analyzed, anticipated and understood. This allows services to be provided, scaled up, downsized or retired to ensure that required service volumes can always be met, while excess capacity does not result in unnecessary costs.
- Where possible, demand from different areas of the business should be aggregated to enable the design and provisioning of standardized services, reducing complexity and cost.
- The impact of changes in demand is quantified and any amendments to the SIAM model necessary to accommodate the changes are reported to the Executive Steering Board and are appropriately authorized and resourced.

Changes in demand

There can be changes in demand owing to seasonal activity or events in the business cycle.

The food industry may need different SLAs around harvest time, whereas a manufacturer may require more stringent controls in the time leading up to a product launch.

Governance controls for demand management should include:

- Defined process and policies for demand management.
- Roles and responsibilities clearly defined and allocated for demand management, in all SIAM layers.
- Adequate resources allocated to performing demand management activities.
- Countermeasures to prevent business units bypassing the governance framework and creating their own 'shadow' arrangements.

- Regularly reviewed metrics and measures to monitor the performance of the demand management capability and drive continual improvement.

2.4 Define principles and policies for roles and responsibilities

As part of the Discovery & Strategy activities, the SIAM project must define the principles and policies that will guide the definition of roles and responsibilities during the Plan & Build stage (see section **3.1.6 Detailed roles and responsibilities)**.

It is recommended that a baseline inventory is created so that there is a clear means to address the question, *"What skills do we have?"*.

Many organizations fail to define roles and supporting skills clearly, resulting in individuals being asked to perform tasks beyond their ability. Conversely, organizations may include individuals who have skills that they are not using. Creating role profiles first and assessing people against them, risks missing skills that already exist in the organization but are not being used or maintained.

It is crucial to define principles and policies for roles and responsibilities for the customer organization (including the retained capabilities), the service integrator and the service providers within the planned SIAM model. Once role profiles are created, they can be compared against available skills and plans made to develop or procure the required skills and experience.

It can be useful to use a standard framework when defining the skills for each role, as this helps to ensure a consistent understanding across all layers. Use of a 'common language' for describing skills is also beneficial, compared to any proprietary skill definition, which may only have relevance and understanding within one and not all parties within the SIAM environment. It is important to be clear and unambiguous about the skills and level of experience required when recruiting, whether that be for permanent staff, service providers, contractors or temporary staff.

Skills frameworks

The Skills Framework for the Information Age (SFIA®)

SFIA provides a common language for the skills and competencies related to information and communication technologies, digital, software engineering, cyber security and other technology-related roles.

SFIA is used in a number of ways, including for individual assessment to confirm the current skills inventory, definition of role profiles and job descriptions, identification of skills for projects and other change initiatives, gap analysis and development action planning to help individuals and their organizations.

Every role in a model requires a specific set of skills. Defining these and ensuring people have the right skills for each role increases the likelihood of success. This applies equally to internal and external staff (such as those from external service providers or contractors).

Japanese iCompetency Dictionary (iCD)

There are several locally developed frameworks in existence, for example the Japanese iCD. The Japanese IT Promotion Agency (IPA) that produces the iCD, has collaborated with the SFIA Foundation to map iCD to the SFIA framework.

e-Competence Framework (e-CF)

The European e-Competence Framework (e-CF) provides a reference of 40 competencies as applied at the information and communication technology (ICT) workplace, using a common language for competencies, skills, knowledge and proficiency levels that can be understood across Europe.

The following principles for defining roles and responsibilities are useful when designing the SIAM strategy and outline SIAM model. They should be used to support the definition of roles and responsibilities in the Plan & Build stage:

- Roles and responsibilities must support the goals, objectives and vision for the SIAM ecosystem, as defined in the SIAM strategy, outline and full business cases.
- All definitions must be relevant to the SIAM model. Generic definitions from another model can be a useful starting point, but must be reviewed against the design for the ecosystem to ensure relevance.
- All defined roles and responsibilities must include necessary capabilities, skills, competencies, knowledge and experience.
- There should be a set of distinct definitions for the roles within the customer organization, retained capabilities, the service integrator and the service providers with a clear demarcation.
- Definitions should recognize that the service integrator is the customer organization's 'agent' and, as such, takes a governing position within the ecosystem.
- Definitions should include integration and collaboration responsibilities where appropriate.
- Roles and responsibilities must cover all stages of the SIAM roadmap, recognizing that different skills and competencies will be required at each stage.
- Where there are existing contracts that will continue to be used in the new SIAM model, any roles and responsibilities defined in them must be reviewed and, if necessary, updated.
- Clarity of roles, responsibilities and ownership is especially important when the model is operated over multiple geographies, locations (sites) and organizations. In these complex environments, there is a higher risk of roles being duplicated or not being filled.
- It is advisable to use a consistent skills framework.
- When a skills framework is used, focus on SIAM role requirements and use appropriate skills and levels to describe what is needed.
- Resources, roles and responsibilities are required to run existing services as well as developing and implementing new ways of working. It is often necessary to create role profiles for initial implementation and transition that are different from those required for ongoing operation of the SIAM model.
- When defining roles and responsibilities, consideration should be made to any changes affecting existing staff (see section **3.2 Organizational change management approach**).

- Requirements for segregation of duties should be considered at all layers and for all roles (see section **2.3.9 Segregation of duties**).
- The overall structure and division of responsibilities should be subject to ongoing audits, review and continual improvement, to ensure roles are conducted as defined and continue to be appropriate – reflecting changing business and customer needs.

When individuals are asked to take on roles in addition to their normal 'day job', it is necessary to assess whether they have the required level of authority, autonomy, influence, capacity, skills and experience to carry out what is required successfully.

2.5 Analyze the current state

It is vital to understand the current and expected future state of the organization's internal and external environment before the SIAM strategy and the outline business case can be created, and the outline SIAM model is designed.

2.5.1 Mapping the landscape

Mapping the landscape to create a baseline assessment provides information on the situation the SIAM transformation aims to change (the 'as-is' position). It provides a critical reference point for assessing changes and their impact, as it establishes a basis for comparing the situation before and after a change, and for checking the effectiveness of the project. Baseline information should be collected in such a way that the same type of data can be reacquired after the project, to compare the results and assess the extent, or lack of change achieved.

Sources of information for baseline assessments include:

- Observation
- Interview
- Analysis of existing documentation, including contracts, process descriptions and role descriptions
- Analysis of service performance reports
- Analysis of existing artefacts, such as toolsets, service catalogs, etc.
- Formative situation analysis
- Stakeholder analysis
- Resource mapping
- Service management maturity assessment
- Process maturity assessments
- Technology assessments
- People capability assessments

These determine the starting position that can be used to identify any gaps in capability and resource. When mapping the landscape of the current service model, the following areas are important to establish and communicate:

- Current services
- Current service providers (internal and external)
- Stakeholders
- Customer attitudes to the service, including satisfied and unsatisfied needs
- Stakeholder objectives and assessment of their position (influence and favor)
- Current service performance
- Transaction volumes
- Technical landscape – including out of maintenance components, vulnerable elements, licencing and standardization
- Assets, configuration and architecture
- Existing contractual positions, including obligations and tenure
- Contractual rigidity and incentives
- Service provider and supplier cost modeling and flexibility
- Commercial, technical, service fit and obligations
- Financial model and budget
- Process, service and technical interfaces, interactions and dependencies, identifying broken links and single points of failure
- Service components that are due to change and when, including critical dates
- Flexibility – existing and required
- Existing personnel, their skills and capabilities
- Staff strengths, weaknesses, performance, aspirations and vacancies
- Critical or hard to obtain skills
- Jobs market and the ease by which resources may be recruited, disengaged or retained
- Any 'known unknowns' or upcoming change

It is useful to express these in terms of a statement of requirements (SoR) and a 'data room'.

Data room

A data room is a place where all information regarding an organization or situation is stored, such as a library, where people can go to learn about the organization or situation and check facts and data. The types of data that can be stored are not limited, but examples of what could be included are:

- Organizational structure and business operating model
- Business strategy documents
- IT strategy documents
- Operating model and structure
- Charging model, including whether IT runs as a not-for-profit model, cost model or profit model
- Service catalog(s) confirming the volume of services and their users
- Service reports (ideally for a minimum of 13 months) to provide information on trends in performance
- Contracts – who supports IT, how and for what?
- HR information on number of employees and how they are split across the organization
- Risk registers, issues registers
- Known debt/losses
- Legal situation

This list is not exhaustive but illustrates that the types of data held in a data room can be as varied and as inclusive as necessary. Data rooms often support commercial activities to allow lawyers, analysts and procurement professionals to interrogate the data for information.

A data room can be physical or virtual. This will depend on the preference of the stakeholders involved. The data room will be made available during the due diligence period of contract negotiations, but is often populated as soon as the decision has been made to go to market. Access to it will be restricted as it often contains highly sensitive information. Non-disclosure agreements must be signed by all parties.

2.5.2 Analyze the marketplace

Mapping the landscape produces a picture of the services currently provided to the organization. It is also important to look externally and establish the commonly encountered market offerings outside the organization. This includes assessing what is offered, by which service providers, to organizations of a comparative size.

This analysis should include assessing potential external service providers and external service integrators, and the SIAM models in which they operate. Organizations can look outside their industry and geographic area, but they must realize that different drivers can affect choices made by organizations in those sectors.

For physical IT products, many offerings are fairly standard and comparable, for example, data center hosting and commodity cloud services. Other services may be more flexible or even include combined standard and non-standard offerings, for example, standard applications with customized levels of application integration, coupled with support and hosting.

The same is true for external service integrators. Most have a standard core offering with additional options and many also offer customization.

Caution should be taken in deviating too far from the norm. Customization creates variation and this can result in additional costs and complexity that outweigh the value achieved. A standard offering can be more easily replaced than a non-standard one, often referred to as 'loose coupling'.

The customer organization should establish the variables in the sourcing model:

- Which services are commonly combined (for example, end-user computing and service desk) and why?
- Which services are commonly offered as standard services (for example, cloud services) and which are customizable?

Many service providers set a minimum viable size of a contract, often expressed as annual contracted value. If the contract structure that the customer organization is considering implies a smaller value than this, then the potential service provider may suggest combining services to increase the value of the contract or may decide not to bid.

As the customer organization formalizes its strategy and defines what it would like to invite service providers to bid for, a good approach is to test the plans informally with a sample of service providers. If this is carried out before proceeding further into the decision-making stage, and honest feedback is received from trusted and established service providers, this may avoid issues later on.

Open discussion to evolve understanding

A public-sector organization planned to adopt SIAM but wanted to explore possible approaches before finalizing its SIAM strategy. It used a government marketplace tool to find potential external service integrators. It was invited to attend an open discussion on the service integration marketplace and the customer's objectives.

The customer provided information on its services, internal capabilities and desired outcomes from a SIAM implementation, and invited discussion. The service integrators gave their views and offered advice on potential SIAM models that might be appropriate for this customer.

This helped the customer to understand what its options were and to confirm its requirements before launching formal procurement notices.

Well-regarded service providers are likely to have more opportunities than they can pursue. They may be selective about the opportunities for which they bid, as the process can require significant investment in time and money, can last many months and require a strong team. Large service providers tend to prefer working with large customers. A small customer, or a small request, might not be attractive to large service providers, and it is more likely to obtain better responses from smaller service providers.

A deluded buyer

A small organization within the UK health service needed to renew its IT services across a commonly encountered set of services. It decided to adopt a SIAM model and defined six service 'lots'.

After many months' effort and much expenditure preparing the procurement documentation, and ensuring conformity to regulatory requirements for government procurement, it went to market.

The product of its initial market engagement was a procurement charge of £1.2m, two vaguely interested service providers and no bidders at all for one of its six lots. The procurement failed.

2.5.3 The role of analysts, benchmarkers and advisors

Some organizations like to validate a decision with research before finally making a commitment. Analysts have a role in this validation, but it is important to recognize that they often make their money by selling research to customer organizations as well as to service providers. The best analysts can accelerate a market survey and support their assertions with facts. The worst are partisan or biased, and may recycle what they have been paid by a service provider to develop, repackaging it for a customer.

Benchmarkers collect contract data and construct models that are designed to provide evidence-based price comparisons. This data can be useful in supporting 'what-if' analysis rapidly, to give an indication of what each option would be likely to cost in the market. Such data is a proxy for market data, being much quicker and cheaper to obtain than fully worked out quotations. Benchmark data is also used in longer agreements to ensure that the price quoted is in line with the current market price.

Advisors are often hired by organizations to support them in the design and implementation of the SIAM model. Customer organizations must check that the advisors have knowledge of a broad range of SIAM models, their procurement and implementation, and that they do not favor one model or service integrator. By being experienced in the process and aware of the leading service providers, advisors can bring rigor and methods that accelerate the process for a customer. This can reduce the costs of sale for the service providers. Advisors can also draw on and apply analysis and benchmark data.

Some external commercial organizations combine analysis, benchmarking and advisory services. It is good practice to get independent advice as organizations that are also service integrators as service providers may favor their own SIAM model over others.

2.5.4 Current organizational capability

Capability

"The power or ability to do something."[15]

---

[15] *www.lexico.com/definition/capability*.

> Capability includes people, processes and tooling.

To support the design of the SIAM model and the decision about which SIAM structure to adopt, it is necessary to assess the organization's current state and existing capabilities to confirm its maturity levels. This exercise will help the organization to identify the capabilities in its current operating model that are not at the desired level of maturity, and whether it has the capability or the appetite to address the shortfall. Organizations often use an external adviser to support their maturity assessment and to provide an objective view of their capabilities.

This assessment should also be conducted as part of the later roadmap stages by potential service integrators and service providers that would become part of the SIAM ecosystem. It is recommended that all organizations assess their staff's skills and use this as a baseline for evaluating the requirements and risks associated with the project, operation and the underlying business case and benefits.

Baseline capability assessments should be carried out before the transition project commences. These help to understand the 'as-is' situation and form the baseline for the desired future 'to-be' situation.

> **An honest assessment**
>
> Unfortunately, many organizations ignore assessment activities to save time or resources.
>
> Other organizations conduct internal analysis and create outputs that are wildly different from reality. Organizations may be overly optimistic, so they receive a surprise during the Plan & Build stage when the reality does not match the assessment outcome and the required capabilities are missing.

The capability assessment consists of two major phases:

1. **Creating a current state capability portfolio**
   The first step is to identify the organization's current capabilities for the SIAM model. These can be a combination of service management processes, operating functions, practices, project and program methodologies, and supporting tooling and technologies. The obvious starting point is the current operating model and responsible, accountable, consulted, informed (RACI) matrices for processes. Every capability currently provided within each function and team should be identified and mapped to the operating model.

2. **Capability assessment**
   The current state capability portfolio should be compared with the SIAM vision and objectives and the target SIAM model. An objective of the capability assessment is to identify how well the current capability portfolio supports the SIAM strategy. Recommendations should be made about the capabilities to be retained within the service integrator layer, which to mature within the customer organization and those for which an external service provider should be chosen. This will help to identify which SIAM structure is most appropriate. For example, good capabilities might lead to an organization choosing an internal service integrator rather than an external service integrator or a hybrid structure.

2.5.4.1 Assessments

It is important to agree the scope of any assessment, for example, the processes and functions to be included, which documentation is to be reviewed and the methods to be applied. It is also necessary to agree if the assessment is to be managed internally by a quality team or internal auditor or if an independent assessor should be used.

There are several approaches to performing an assessment, including individual or group stakeholder interviews, or process walkthrough workshops with process owners and managers. Assessments should be based on industry best-practice frameworks and standards, such as ITIL or SFIA, and should assess:

- Process adoption and consistency across the organization
- Business alignment and value
- Process metrics and measurements
- Supporting service management tooling
- People, including organization, roles and skills
- Current levels of integration for people, processes and tools
- Levels of continual improvement for processes and functions

Note that these assessments need to focus on the SIAM practices for processes, capabilities, roles, etc. and not 'traditional' service management maturity.

Example inputs to the assessments will be:

- Existing capability model
- Stakeholder maps – including process/capability owners and managers
- Industry frameworks and standards
- Current processes and procedures
- Current RACI matrices
- Tooling architecture
- Process touch points and interfaces

The expected outputs from the capability assessments are:

- Current state assessment
- Capability heat map
- Individual capability maturity levels
- Retained organization training and development plans
- Individual process maturity levels
- The feasibility of addressing improvements or gaps with customer organization resources
- Tooling recommendations
- Draft future state model

Heat map

A heat map is a graphical representation of data where the individual values contained in a matrix are represented as colors. Heat maps are often used to summarize findings ranging from areas that require attention ('hot') to those that are well established/stable/mature ('cold').

2.5.4.2 Operational benchmarks

Health checks and maturity model assessments are forms of operational benchmarking in which the quality of an aspect of a service is compared to external norms in a structured manner. Some of this data is statistically valid and some is supported by benchmark data obtained from a variety of comparative organizations.

The purpose of conducting these assessments is to check whether the area under review functions as it should, to identify corrective actions or improvement opportunities, and to measure its effect in achieving the expected results. Typically, these actions can be managed through continual improvement activities. However, if any findings have the potential to affect the success of the SIAM model or the transformation, these will need to be addressed through the transition project. In the context of sourcing, it helps to consider checks and maturity models, such as:

- CMMI for Service (CMMI-SVC): administered by ISACA, provides an approach but not comparative data for service management. Can be independently assessed for certification purposes.
- SFIA: a framework of internationally recognized descriptions for responsibility characteristics and skills practiced at various levels of responsibility and experience by people working in IT-related roles.
- IACCM (International Association for Contract and Commercial Management): benchmark contract management process performance.
- One of many ITIL maturity assessments or the COBIT Process Assessment Model (PAM).

There are also frameworks to which current practice may be compared. Useful examples include:

- Service management: ITIL
- Project management: PRINCE2, PMBoK or the Association for Project Management (APM)
- Program management: Managing Successful Programs®
- Architecture: TOGAF®
- Governance: COBIT

These are non-prescriptive (as opposed to standards) and may be adopted and adapted to meet specific requirements. This is important, as these models, assessments, frameworks, practices, benchmarks and approaches are not necessarily tailored to the specific requirements of a SIAM model and so need to be adopted, used and interpreted.

2.5.4.3 Process maturity

The majority of SIAM transition projects do not start from a 'greenfield' situation. More regularly, a transformation is proposed to change one or more existing delivery models, where service providers, internal resources and services are already in place. The SIAM model and supporting processes may, in part, be based on the existing processes of current service providers and internal capabilities, or may have to operate within a wider enterprise process framework (EPF). Regardless of whether the proposed processes are entirely new or not, any parties undergoing transition to the model need to understand the starting point.

A process maturity assessment may be required, particularly if the future model is built upon or developed from an existing model. A maturity assessment can provide a baseline of the current process capabilities across the organization. It will support identification of existing good practice or weaknesses, both across the overall organization and within each part of the organization. For instance, CMMI® for Service provides a well-framed assessment, although it is more for traditional service delivery processes, not necessarily those applicable in a SIAM model.

Limited timeframes

Many SIAM projects have limited time to execute the Discovery & Strategy, and Plan & Build stages of the roadmap. If this is the case, the customer organization might have to compromise on the time spent on the current state assessment. A more pragmatic approach may be to 'go with what is there' and make changes as the project unfolds. However, always ensure appropriate levels of control on such changes.

A full assessment is important in order to drive an improvement or transformation program, but it may have to be carried out in tandem with operational responsibilities.

A process maturity assessment provides:

- Valuable input into the design of the future process model
- Opportunities for reuse of existing, known and proven processes
- Information about whether processes need tailoring or complete redesign

The assessment outputs will provide an understanding that is not only used during the process design activities, but also builds support among those parties who must later implement and operate the new or tailored processes.

Process models define activities, roles and responsibilities required to effectively execute the process activities. The RACI matrix is a relatively straightforward tool that can be used for identifying roles and responsibilities during a SIAM transition.

RACI

RACI stands for Responsible, Accountable, Consulted and Informed. These are the four key 'involvements' that can be assigned to an activity and a role.

A RACI chart or matrix shows all the activities or decision-making authorities in an organization set against all the people or roles.

2.5.4.4 References

Organizations may also speak to similar organizations that have adopted SIAM to collect their own reference data for SIAM and for prospective service providers and/or external service integrators. It is important to collect as much information as possible to support a decision to transform to a SIAM model. Conferences, online forums and social media can also be used to collect information about other organizations' SIAM experiences.

2.5.5 Existing services and service groupings

Services, service components and service boundaries change, and the current services should be defined first. Following this, their suitability for current and expected business conditions is assessed. This can be a challenge as services can underpin, support or enhance other services, so the boundaries may be blurred.

2.5.5.1 Service definition

The aim of service definition is to understand and map all discrete services and their dependencies in a hierarchy, and define which service provider is responsible for each service or group of services. The definition should include:

- Current service definitions (whether provided internally or externally).
- Service boundaries.
- Service providers for each service and their geographical location, which includes internal operational support units or subcontractors:
  - Services that are sourced from a single service provider (a service group).
  - Services that are sourced from multiple service providers, for example, applications development.
- Contract terms, including terms that need to change.
- Contract expiration, with length of time left to run, implications of early termination and exit clauses.
- Agreed or contracted service levels.
- Performance against contracted service levels, usually at least 13 months' reports and transaction data where possible (which equates to one year, plus the same month last year). This provides a snapshot and a trend.
- Service user and senior stakeholder attitudes to the services and service providers.
- Requirements that need to change. These may be derived from the analysis of which current obligations are working effectively and where change is required (See section **2.5.6.1 Assessing contracts**).
- Desired service levels, including which targets are essential and whether they are currently covered:
  - If higher service levels are requested, is there funding or sponsorship for the difference?
- Seasonal variations and peak demand, key business events and requirements.
- Identified integration or handoff issues between the components or service providers that a new SIAM model would need to address. What issues cause friction?
- Data on assets, people, services and transaction volumes.

- Asset ownership.
- Knowledge ownership.
- Costs, charging models, external service provider costs, budget.
- Related services (for example, project work, catalog items, variable elements).
- Changes planned and scheduled.
- Critical events that act as a constraint (can anything be done about them?).
- External service provider support and maintenance contracts.
- Hours of cover, with a possible specification of core versus extended hours.
- Languages and other parameters covered, for example, by each operational unit or by each service.
- Software licence agreements.

Note that the gathering of information about the services is not confined to the retained capabilities, although it is likely to start there. It must include all other business areas and stakeholders, for example:

- Finance
- Commercial
- Sponsors and executives within business units using the services
- Service users
- Project managers
- Front-line business operations

These stakeholders may not only have information on the existing services, but they may have needs that are unfulfilled and that should be factored into the definition of the desired future state, if they are economically viable.

In some cases, business units may have unfeasible hopes that are quite uneconomic to fulfil. If presented with an expensive requirement, consider asking '*How would we persuade the finance director of the value of this and who would pay for it?*'. During requirements gathering, avoid making any commitments beyond a promise to consider what is requested.

Perceived service levels

As part of a transition to a SIAM model, an organization wanted a new wide area network (WAN) service. As customer satisfaction was good, the contract team included the service levels that were in the contract with their existing service provider as part of the specifications.

Thankfully, when the customer's in-house service management team reviewed the schedules, they were able to point out that the service levels received were significantly better than the ones documented in the contract.

2.5.5.2 Service mapping

Understanding the full range of services can be a challenge, as many services can underpin, support or enhance other services, so the boundaries may not be clear. A good approach is to use service catalog techniques described in other practices (such as ITIL). Some organizations may already have a comprehensive service catalog that will provide the necessary information.

At the lowest level of the hierarchy you are likely to find 'infrastructure' services such as:

- Data centre hosting
- Local area networks (LANs)
- Wide area networks (WANs)
- Telephony

At the highest level will be the application services that are used directly by the customer organization's users, such as:

- Email
- Word processing
- Order management
- Payroll
- Service desk tool

There may also be technology services that are used to support the applications, such as:

- Internet services
- SMS text services
- Mobile phones

And non-technology supporting services, such as:

- Desktop support
- Service desk
- Payroll support provided by an external service provider
- Facilities management

Useful sources of information include:

- Service desk tools
- Discovery tools
- Technical architecture models
- Configuration management databases
- Support contracts
- Budgets
- Individuals' knowledge of IT, service management and the business

2.5.5.3 Intellectual property

Some of the information gathered will be important to the procurement strategy but may need to be kept confidential. Consider what needs to be shared with whom, when and how.

Some information will be commercially sensitive and may be the intellectual property (IP) of an incumbent service provider. In this case, ensure that it grants permission to share this. It may be necessary to create new documents containing appropriate summaries that do not breach any duties of confidence.

It will be imperative that any future contracts contain clear clauses about the use and ownership of IP for all parties. Understanding those considerations at this stage will help in the Plan & Build stage later (see section **3.1.3.8 Intellectual property**).

2.5.6 *Existing service providers*

It is often difficult to dissociate services from their service providers. Consider:

- What are the attitudes to each service provider, together with those of operational, senior IT and service users? Which would be favorable to keep?
- What activities are delivered consistently well, badly or have problems?
- How often must recourse be made to the existing contract? The relationship is probably better if contracts are consulted rarely, compared to when the parties are constantly referring to contractual obligations to determine if they are operating the service correctly – or not.
- How well do the service providers collaborate with each other?
- How effective are relationships with/between service providers? Is it 'us and them'?
- How much effort, time or money is spent with each service provider?
- How well do their service contracts align to current business needs?
- What are the key breakpoints and termination dates? How easy would it be to change them?
- Would the incumbent service provider want to continue supplying the services? If so, what would they want to change?
- What is the intellectual property (IP) position? Is it easy to change service provider and keep the service information?
- How could changes to the service provider and supply contracts be managed from the current arrangement to the future SIAM model's needs?
  Why would each provider want to agree to the proposed change (remember that changes must be agreed by both parties)?

Role playing

Consider the use of different teams, playing different roles, to work through scenarios to anticipate incumbent service providers' reactions to potential changes to test which are achievable. This uses staff from the customer's own organization in role play, to test a strategy in private before going public with it. The roles are:

Team 1 (sometimes known as 'red team'): colleagues who are not involved but have relevant skills and experience. They pretend to be the counterparty, in this case the service provider (or if you are a service provider, they will play the role of the customer).

Team 2 (sometimes known as 'blue team'): these are team members proposing the planned approach.

This approach can be used to quickly discover weaknesses in a plan by using the red team to attack it in private. This can also be used by service providers to assess bids before they are submitted. Here, the red team role plays either the customer or a competing bidder.

2.5.6.1 Assessing contracts

It is useful to assess the state of existing contracts and the way in which service provider performance is managed. Consider the following:

- How detailed are the service definitions?
- Are there service schedules?
- Do the schedules reflect customer needs?
- Is the provider performing consistently?
- Is there a designated point of contact responsible for managing the contract and the relationship with the provider?
- Is the cost model transparent?
- Knowledge of the cost model in use and charging elements such as:
  - Fixed monthly
  - Variable usage (unit-based pricing)
  - By catalog item
  - By project
  - Dynamics of the model and the flexibility to change to the model
- Strength of obligation based on relationship term and criticality of services provided.
- How compelling are the performance incentives?
- Does the contract provide value for money?

If current service provider performance falls short of expectations, and the service provider is to be retained, it is important to understand the root causes and address them. If possible, consider making changes to ensure the contract is aligned to others and is appropriate for the needs of the proposed SIAM model.

2.5.6.2 Incumbent exit and re-bid

Validate that any incumbent service provider exit plans are up to date and available. For any plans that are not sufficient to conduct an exit, issue commercial notice requiring them to be updated so that they are fit for purpose. In some situations, service providers may not be contractually required to have an exit plan. If this is the case, then the customer organization (and service integrator, if appointed) will need to work with the service provider to find an acceptable outcome.

If an incumbent service provider wants to re-bid for services, there needs to be clear communication about how this is governed. There could be the impression that an informal decision has already been made (either in favor of, or against, the incumbent provider), which may discourage other (suitable) providers. In any bid, all bidders should be treated equally and fairly, including the incumbent. An 'ethical wall' may need to be established between the bid management team and the existing service delivery teams, with no sharing of information that is not shared with the other, potential service providers.

2.5.7 The customer organization

The customer organization includes the retained capabilities, possibly elements of the service integrator layer (for an internal or hybrid SIAM structure), and possibly one or more internal service providers. Where customer organization staff have roles in different SIAM layers, segregation of duties between layers must be clear.

The structure of the future SIAM model will be influenced by the strength of the customer organization's capabilities and decisions associated with either retaining or externally sourcing elements of the SIAM ecosystem. The customer organization should consider its own position and how it is perceived by its own market.

Questions such as the examples below are useful discussion points:

- What would an honest service provider say about the organization and what it is like to deal with?
  - o What improvements could be made, before engaging the market and entering a new relationship?
  - o Would an incumbent service provider be likely to re-bid for business with the customer organization?
- What are the interactions and interfaces between current service providers and the customer organization, and how well do they function?
- Are roles and job descriptions accurate and up to date?
  - o Are they an accurate definition of what really happens?
  - o Do the customer organization's capabilities fit well with the roles and capabilities required?
- Is the current tooling fit for purpose?
- Is there a clear understanding of the current costs associated with the service model?
- Is it acceptable to allow, or indeed is it necessary for staff to transfer to or from service providers and the customer organization?

2.5.7.1 Capability and capacity to manage sourcing

The customer organization needs to consider its competence to manage a transition to a SIAM model. This includes time, capability and resources for the associated tasks. These roles are likely to be required:

- Project sponsor
- Sourcing manager
- Current service owner(s)
- Manager for data room preparation and management
- Commercial manager
- Procurement manager
- Financial manager
- Legal support
- Support staff from incumbent service providers
- Project and program managers, including the PMO

Just as a customer organization should assess its ability to fulfil these roles, either internally or externally, a prospective service provider will assess the ability of a customer to deliver the sourcing program successfully.

2.5.7.2 Inflight projects

It is rare for there to be no other changes being made within any of the organizations involved during the period of the SIAM transition. Whether these other change activities are directly linked to the SIAM project or not, consideration needs to be given to the dependencies, constraints and the impact these changes might have on the success of the SIAM project (and vice versa). The SIAM project may support inflight projects, cut across them or be indifferent (see section **3.3.3.7 Accommodating inflight projects**).

These areas must be considered:

- Too much change activity might affect elements of the SIAM project negatively
- The project timelines may need to consider other projects, operational activities or other constraints, especially where resources are shared
- Sponsors of interacting projects may be concerned that their main interest is not interfered with and that any requirements that flow from their projects to the SIAM project are accommodated

In this instance, a PMO is useful (see section **2.2.3.2 Project roles**).

2.5.8 Other influences

When deciding on the SIAM model and sourcing approach, the customer organization, service integrator and service providers must all consider not only their own strengths and weaknesses, but also other factors that will influence their success in the SIAM ecosystem. The following considerations are relevant:

- Specific organizational expectations and requirements
- Customer experience
- Staff experience
- Current service levels
- Information governance and privacy
- Speed and responsiveness to change
- Compliance and standards:
  o Security and data management
  o Legal and industry norms
  o Regulators
- New methods and approaches to delivering services and the presence of new options
- Constraints:
  o What constraints exist?
  o What are the obligations?
  o To whom do they apply? Who can break them and how?
  o What are their effects?
  o Where is their flexibility?

o Are there any unspoken and unwritten rules?

(Note – some may be valid and others just perception or habit)

A multinational organization will have to assess the legal and regulatory framework for each geographic area.

Some of these influences will be stable, while others change over time. All parties will need to factor these influences into their plans. They may be accommodated, overridden or ignored, while each has a risk and value associated with it that needs to be considered.

Local content

It is important to appreciate the factors that will influence a successful contract bid. An Australian government department defined that all tenders must include a local content assessment criterion with a minimum weighting of 30%. Bidders had to demonstrate the local content and benefits of their proposal in areas such as:

- Employment
- Upskilling – including apprenticeships, formal and informal training
- Local industry participation
- Indigenous development
- Regional development

2.5.9 Collaboration

Effective collaboration and teamwork is vital between all parties within a SIAM ecosystem. The different parties must work not only to meet their own (often contracted) requirements, but also to deliver an end-to-end outcome that benefits the customer organization. It is important when mapping the landscape to include aspects concerned with collaboration (see section **3.1.8 Collaboration model**).

Collaboration extends to the current state assessment and the demands for the future. Many organizations focus on the needs and challenges of a single service or service provider independently, and miss the review of collaboration needs and challenges across all areas and stakeholders affected by the SIAM model. Without collaboration, there is a risk that contracts with service providers would be based on delivery targets alone, omitting the requirements for collaboration.

2.6 Define the strategy

Strategic objectives for SIAM

The strategic objectives for SIAM are those that SIAM will deliver in support of the long-term goals of the organization.

2.6.1 What is strategy?

> **Strategy definition**
>
> *"A plan of action designed to achieve a long-term or overall aim."*[16]

Creating a strategy allows an organization to coordinate and plan activities rather than relying purely on opportunity, individual initiative or serendipity. Strategic planning is particularly important in situations where there are many parties to coordinate, where there are long lead times between the development of a competence and realization of benefits, and where there are high-stake decisions that are difficult, expensive or time consuming to reverse.

To provide clarity to stakeholders regarding the overall strategic intent, a concise strategy should include:

- Why the organization must change
- The benefits of change – money, competition, speed, customer service
- What the organization must build or change
- The vision
- How much it will cost – money, people, other
- How long it will take to deliver
- Stages along the way
- Difficulty/risks
- Options: what could the organization do if not this
- At an individual level, an explanation of '*What's in it for me?*' (sometimes referred to as WIIFM)

In a SIAM project, the SIAM strategy should be used to guide all activities in the SIAM roadmap, including design of the SIAM model, sourcing decisions for the layers and as part of onboarding service providers.

> **The importance of strategy**
>
> *"The best way to predict the future is to invent it."*
>
> **Alan Curtis Kay**
>
> *"In preparing for battle, I have found that plans are useless, but planning is indispensable."*
>
> **General Dwight D. Eisenhower**

Within a SIAM model, many parties need to be coordinated and often large investments with long-term commitments are needed. A strategy provides an overarching direction and high-level plan to

---

[16] *www.lexico.com/definition/strategy*.

lead the organization and its people through the transformation. A SIAM strategy must always support the organization's corporate strategy and may even be a component of it.

Different forces influence an organization's strategy. Strategic drivers include elements such as:

- Response to competitive threats from new entrants to a core market
- Aspiration to build a market extension by exploiting an emerging opportunity that is perceived as underserved or poorly served
- Desire to reduce ongoing costs
- Desire to improve the quality of delivered services
- Desire to better structure or source a constantly expanding services department

From these, the organization will develop a strategy to change its form, build new competencies and adjust the allocation of resources. The links between SIAM, corporate and divisional strategies must be clear, understood, documented, agreed and communicated. Decisions made in relation to SIAM (for example, approving the SIAM business case) need to support the corporate strategy. If the organization's senior leaders can see that money and resources allocated to a SIAM project will support the corporate strategy and business outcomes, they will be more likely to approve their allocation because they support the corporate agenda.

2.6.2 Strategic drivers for SIAM

Over time, the number of service providers – as well as the type of services being sourced externally – has increased for many organizations, creating choice and concepts such as 'best of breed'. This has led to increased complexities and management overheads for the customer organization.

Dealing with this large and fragmented service provider landscape can create many issues:

- Service providers operating in silos, focusing only on their own performance targets and not the end-to-end service
- Customer organizations managing and measuring each service provider individually, creating management overhead
- Challenges for the customer organization trying to build an end-to-end view
- Friction caused by fragmented processes, tools and service providers using different ways of working
- Additional overheads and complexities when contracts and associated legalities cause differences in reporting, invoicing, service credits, etc.
- Conflicts between service providers
- Little collaboration between service providers, limiting the efforts for the 'greater good' of the customer organization
- Disaggregated services with no clear accountability

Some organizations believe that consolidating their service provider landscape to one, or a few, large service providers will help to alleviate these challenges.

The SIAM approach differs. It allows a customer organization to benefit from having access to a variety of service providers, while reducing its management overhead. The service integrator provides ownership and coordination across the service providers, in a consistent and transparent manner with clear accountability and responsibilities. A SIAM model makes it easier for a customer organization to onboard and offboard service providers when required.

The strategic forces (internal or external) that influence the transition to a SIAM model are likely to include corporate strategy drivers, as well as those that are more specific to the services that form the SIAM model. Examples are a corporate direction to centralize contracts to gain efficiencies or, from a services perspective, a directive to retire old technology and replace it with more efficient solutions.

Drivers may originate from outside or inside the organization. The SIAM driver types identified in the **SIAM Foundation BoK** include:

- Service satisfaction
- Service landscape
- Operational efficiencies
- External drivers
- Commercial drivers

A good strategy will take account of these applicable driver types, but should not be defined by them alone. Ultimately, a strategy is a plan to reach a goal and advance objectives. It is implemented through tactical and operational plans and activities.

2.6.3 Strategy formation

Strategy should never exist in isolation. It is often expressed in the context of:

- Vision
- Objectives
- Targets

Each of these areas can (and should) be expressed first from the perspective of the organization, and then expanded progressively through each level of the organization.

Strategic planning for SIAM

Strategic planning can be made up of four key steps:

1. Understand the current state ('as-is'), including any issues
2. Describe the future state and how it will address issues ('to-be')
3. Outline the high-level stages to get to the future state
4. Detail the next steps

When carrying out strategic planning for a transformation to SIAM:

- Item 1 is achieved by conducting the activities in the Discovery & Strategy stage
- Item 2 is the SIAM model outlined during Discovery & Strategy, and detailed in the Plan & Build stage
- Item 3 is the full SIAM roadmap
- Item 4 is the detail from the Plan & Build stage

Many organizations now seek to be more agile and responsive. Speed of change can be a success factor if the lead time to strategy alignment and adoption of new services is a concern for the customer organization.

Capabilities required for the formation of a strategy are varied. They normally involve creativity, modeling, scenario analysis, pictures, words and figures. When forming a strategy, it is wise to consult widely with all stakeholders and to examine desired outcomes and options from a variety of viewpoints. A plan that has only considered limited options is unlikely to be robust.

Perspective taking

There is a useful thought process (or cognitive exercise) that can be used in strategy formation. It involves consciously taking a series of perspectives from which to view a situation. Ideally, this is then used in a dynamic mode to work through the choices and preferred selections as a scenario unfolds.

Scenarios can be pursued as if the participants are chess players, considering, 'What would happen if I took the queen's rook?', and playing through the consequences. This involves 'walking a mile in another person's shoes'. Sometimes the experience is painful, but the insight can be invaluable.

There is not one, perfect strategy for any organization. There are however plenty that:

- Are inconsistent: for example, *'We want to complete xxx in six months, but we only have competence to achieve it in 18'*
- Are incoherent: different units are pursuing conflicting objectives
- Focus too narrowly on goals or objectives: strategies are not simply goals, but also include plans to achieve them
- Are improbably risky: under the assumption that it will go right … this time
- Are unattractive to key decision makers: for example, if it affects their bonuses

The art of strategy formation is to consider the options widely and carefully. The resources and budget must then be secured before committing to a course of action.

2.6.3.1 The vision for SIAM

The overall SIAM vision is to build a capability to manage service providers with varied interests seamlessly, to provide a cohesive and transparent service view and meet the customer organization's objectives. This vision will be slightly different for the organizations and service providers in each SIAM ecosystem, but each one is likely to share some common elements. Every party's vision should encourage cooperation, collaboration, trust and a culture of 'win-win'.

From the perspective of a customer organization:

- Build the capability for services to be managed on the customer's behalf, to meet the customer's strategic objectives, providing a cohesive and transparent view of services
- Help the customer to move from a transactional and disjointed way of managing service providers towards strategic partnerships and improved benefits realization
- Bring consistency, efficiency and transparency in the way services are being delivered, no matter which service providers are involved

- Enable rapid adaption to change
- Integrate all aspects of service delivery, including operations, projects, service transitions, performance reporting and continual improvement

From the perspective of an external service integrator:

- Build a continuing positive and profitable relationship with the customer
- Build a continuing positive relationship with the service providers
- Win repeat business, or expand existing scope
- Obtain a referenceable engagement
- Build the capability to manage services on the customer's behalf seamlessly, to meet the customer's strategic objectives, providing a cohesive and transparent view of services
- Help the customer to move from a transactional and disjointed way of managing service providers towards strategic partnership and improved benefits realization
- Help the customer to realize the expected benefits of the services, within the constraints of the contract
- Bring consistency, efficiency and transparency to the way services are being delivered, no matter which service provider(s) are involved
- Enable rapid adaption to change and accelerate transition initiatives with confidence, onboarding and offboarding service providers as needed
- Integrate all aspects of service delivery, including operations, projects, service transitions, performance reporting and continual improvement

From the perspective of an internal service integrator:

- Maintain a continuing, positive relationship with the customer
- Remain as the internal service integrator (protecting its position from external alternatives)
- Build a continuing positive relationship with the service providers
- Build the capability to seamlessly manage services on the customer's behalf, to meet the customer's strategic objectives, providing a cohesive and transparent view of services
- Help the customer to move from a transactional and disjointed way of managing service providers towards strategic partnership and improved benefits realization
- Help the customer to realize the expected benefits of the services
- Bring consistency, efficiency and transparency in the way services are being delivered, no matter which service provider or providers are involved
- Enable rapid adaption to change, and accelerate transition initiatives with confidence, onboarding and offboarding service providers as needed
- Integrate all aspects of service delivery, including operations, projects, service transitions, performance reporting and continual improvement

From the perspective of an external service provider:

- Build a continuing positive and profitable relationship with the customer
- Develop a continuing positive relationship with the service integrator
- Maintain a continuing positive relationship with other service providers
- Win repeat business or expand the existing scope
- Obtain a referenceable engagement
- Deliver services to the customer to meet its requirements
- Help the customer to realize the expected benefits of the services within the constraints of the contract
- Provide services consistently, efficiently and transparently
- Adapt to change rapidly

From the perspective of an internal service provider:

- Maintain a continuing positive relationship with the customer
- Build a continuing positive relationship with the service integrator
- Remain as an internal service provider (protect its position from external alternatives)
- Build a continuing positive relationship with other service providers
- Deliver services to the customer to meet its requirements
- Help the customer to realize the expected benefits of the services
- Provide services consistently, efficiently and transparently
- Adapt to change rapidly

2.6.4 Communicating the strategy

A strategy is only as good as the action it supports and enables. Action can be encouraged by communicating the strategy to others, so that they support it. There are three main forms of strategic communication, all of which must take place as soon as possible after the strategy has been approved:

1. Face-to-face communications from the sponsors and leaders of the project, inspiring and promoting the change. This should be persuasive and engaging, and needs to be repeated and referred to regularly, otherwise it may turn into '*something someone once said*' and be forgotten or ignored.
2. The documented strategy itself, which needs to be available, understandable, compelling, as concise as possible, and explain where each party fits into the strategy.
3. A business case that expresses the approach, the outcomes, the resources, risks, approach and cost. This is essentially thoughtful and analytical, with emotional elements.

When implementing a SIAM model, the storytelling and business case precede the transition. This means that by the time of their appointment, incoming service providers fully understand what is expected and how they fit in.

Strategy launch meeting

At the end of the Discovery & Strategy stage of a SIAM transformation in a small organization, all the employees were invited to attend a launch event. This was led by the chief executive, who outlined the vision for the future. He was supported by the other members of the board, who each explained a particular element of the strategy. The event was recorded and made available on the organization's intranet.

Once the external providers were appointed, a similar event was held, but this time attended by the chief executive and senior managers from each of the providers. After the chief executive went through their strategy, the executives from each provider talked about how they were going to support achievement of this strategy.

One potential challenge can be the degree of inertia that exists within an organization. Many organizations have no experience of making a transformational change as significant as SIAM, and many people are not motivated to change (believing that it is 'better the devil you know'). A SIAM strategy will simply not get started or be sustained without a sufficiently powerful reason to act – and to act now. A perceived or actual crisis (positive or negative) can be utilized to show the organization that remaining in the current state is intolerable. Additionally, using stories, hopes and fears may overcome this inertia.

What happens if we do not?

The first of J.P. Kotter's *The 8-Step Process for Leading Change* is to '**create a sense of urgency**'. [17] This requires the organization's leaders to explain why they are planning to embark on a particular strategy, which helps others see the need for change through a bold, aspirational opportunity statement that communicates the importance of acting immediately.

As an alternative, try reversing the statement and pose the question: *'What happens if we do not?'*.

A positive crisis

A public-sector organization had been created from several mergers over time. This resulted in four different IT departments, each with its own heads of IT, tools and ways of working. In addition, they had ten external service providers supplying managed IT services, but with no single point of management and coordination.

Two events made the need to change very apparent:

- One of the external providers failed its availability service level consistently. This prevented the organization from achieving targets set by the government.

---

[17] *www.kotterinc.com/8-steps-process-for-leading-change/*.

- There was a major outage with a critical business service provided by one of the internal IT departments. This delayed the provision of key information to the government.

These events were used as part of the justification for a strategy to move to a consistent SIAM model, embracing both internal and external providers.

2.6.5 Stakeholders

Resistance to a strategy can be overcome by creating awareness and gaining support from stakeholders. Stakeholders are those with an interest or concern in something. They have influence (to varying degrees) on resource allocation and decisions. When producing a strategy to transform to a SIAM model, consider the perspectives of these stakeholders, their likely reaction to the proposed approach and the effect of these on the transition plan, including:

- Customers – both existing and potential – of the customer organization
- Customer organization – senior executives, notably the SIAM sponsor
- Champions and influencers – prominent figures in the organization with an interest in the project but no direct involvement
- End users within the customer organization
- Customer organization – delivery staff, including internal service providers
- Consulting and specialist organizations that may be brought in to assist with the SIAM transition project
- Funding agencies
- Any functions and projects competing for resources that might instead be assigned to SIAM activities
- Regulators
- Incumbent service providers (existing contracts, obligations and their commercial strength)
- Potential service providers, including any potential external service integrator
- Competitors in the market

This list is not exhaustive, and the stakeholders will be different for each organization.

Stakeholder characteristics include:

- Degree of influence
- Degree of power
- Association and relationship
- Trust
- Reliability
- Support
- Attitude
- Interest, priorities
- Hopes, fears and aspirations

These characteristics may make them positively, neutrally or negatively disposed to the SIAM model. When forming a strategy for SIAM, it is helpful to identify those with power or influence over the outcomes (see section **2.5.1 Mapping the landscape**).

Stakeholder buy-in can be achieved by creating a strategy that:

- Achieves the best overall value from the perspective of those with the influence to determine its ability to succeed
- Accommodates interests and priorities to the greatest degree possible (and explains where they are not)
- Is supported throughout the duration of the program and the stages of the roadmap

---

'Selling' the strategy

Techniques for stakeholder mapping, which can be useful in gaining stakeholder buy-in for a new strategy, can often be found in the sales community. Major account salespeople have developed approaches for mapping relationships and visualizing connections to develop strategies for dealing with them and winning approval for new business.

These techniques can work equally well for selling a SIAM model across stakeholders.

Not every individual will establish positive relationships with everyone. Consideration needs to be given to all members of the team when assessing who is best placed to engage with stakeholders. Success depends on trust and respect, and although 'liking' an individual is helpful, it is not essential.

Stakeholder mapping should then be used for stakeholder management activities throughout the lifecycle, as part of organizational change management (OCM) (see section **3.2 Organizational change management approach**). The relationships built at the initial stages of the transformation to a SIAM model will also help to maintain it once it is implemented, so it is essential to consider the individuals involved.

2.6.6 Managing strategy

The transition to a SIAM model involves changes that will affect many elements of the organizations involved. It requires resources to deliver it and sponsorship, won through the construction of a business case (see section **2.7 Create an outline business case**).

A strategy must be seen to deliver change. The right to continue comes from the delivery of meaningful results that generate credibility. This allows the sponsor to sustain the project through various (and sometimes iterative) stages. If the SIAM project is not seen to deliver change, it may drop down the management agenda and resources, and focus will switch to other projects. The strategy should be reviewed regularly and, if necessary, changed.

Losing strategic focus

Many organizations have started the journey of transforming to a SIAM model, developing the strategy, the business case and initiating the program. However, if not actively managed, the attention of the senior stakeholders can move to newer and more exciting areas, such as new business opportunities or new digital solutions.

> This means losing the long-term commitment for the SIAM transformation and can mark the beginning of the end for the SIAM transformation program and its ability to fully deliver the SIAM model and identified benefits.

Impractical or unachievable strategies need to be identified quickly and reassessed. A common reason for failure is organizations that try to move rapidly on too many fronts with insufficient resources.

2.7 Create an outline business case

A clear and understandable outline business case will clarify the purpose of designing and implementing a SIAM model, the expected costs and outcomes. This needs to be a reference document that can be used in the Plan & Build stage to create the full business case, to use in the further roadmap stages.

The business case needs to contain:

- **Strategy for SIAM**: this needs to outline the strategic objectives for the SIAM model, whether the intentions are to address specific pain areas, financial considerations, or operational and business requirements.
- **Proposed services**: the proposed services to be provided.
- **Outline SIAM model**: a high-level outline of the intended SIAM model.
- **Current operating model**: a high-level representation of the current delivery model, including current pain areas that have led to the need for change.
- **Benefits expectation**: the expected benefits from transitioning to the SIAM model, including defined benefits measurements along with current and target values for each measure.
- **Costs**: an estimate of the costs involved in the transition to the SIAM model, with the required level of detail in terms of cost types, cash flow projections and other financial techniques (often specific and established within the organization). Also, estimated total cost of ongoing operation, which can then be compared to current costs.
- **High-level plan**: an outline of the transition to the SIAM model with broad timelines and activities listed along with major deadlines.
- **Risks**: the risks of implementing (and those of NOT implementing) the SIAM model, at a high level. The risks should be classified according to the organization and delivery units affected. This may include risks to the business, risks to the IT organization, risks to service providers and risks to the SIAM environment or ecosystem.
- **Critical success factors (CSF)s** (see section **2.7.2 Critical success factors**).
- **Analysis**: a documented analysis on the viability of the business case (perhaps using the pain value analysis or cost benefits analysis methodologies).

2.7.1 Create the outline SIAM model

The outline SIAM model uses the information and outputs from previous activities in this stage, and in conjunction with the SIAM strategy, can form the basis of the outline business case. The outline SIAM model establishes how it aligns with the intended objectives and enables achievement of those objectives.

The outline SIAM model should include:

- Principles and policies
- Governance framework
- Outline roles and responsibilities
- Outline of process models, practices and structural elements
- Outline of services and service model
- Service providers to be retired

The outline SIAM model will be used in the Plan & Build stage to create the detailed SIAM model (see section **3.1 Design the detailed SIAM model**).

2.7.2 Critical success factors

Critical success factor (CSF) is a management term for an element that is necessary for an organization or project to achieve its mission. It is a critical factor or activity required for ensuring the success of an organization.

CSFs for a successful SIAM model include:

- Well-conceived SIAM strategy
- Appropriate SIAM model design
- Managing process outcomes, not activities
- IT as a strategic partner
- Impartiality of the customer retained capabilities and the service integrator
- Clear boundaries of responsibility
- Governance model
- Confidence in the competence of the service integrator
- Trusted relationships
- Service agreements
- Management information and reporting the right data

The CSFs will provide an input into the outline business case, which is described further in the next section. The challenge is that it is not known whether the strategy and design are well conceived until they are implemented. Sometimes, the approach appears sensible on paper but fails because the flaws become exposed only at the implementation stage.

2.7.2.1 Well-conceived SIAM strategy

If the customer organization is not clear about its objectives and the value it wishes to receive, any associated policies and strategies will also be unclear, increasing the delivery challenges for the service integrator and the service providers.

2.7.2.2 Well-conceived design of the SIAM model

To achieve the desired outcomes, the full SIAM model, including the service model and sourcing approach, should be agreed by the services customer. The customer organization must make decisions on how services will be sourced and which type of service providers, under which structure, will facilitate delivery. The customer organization must ensure that the models are scalable, so that new business processes and functions can easily be added, extended or discontinued.

When appointed, the service integrator will become involved, and the final model should be a joint enterprise. If the service integrator has more experience, the customer should leverage this rather than imposing its own (possibly flawed) model.

The design elements of SIAM incorporate the concepts of separation of concerns, modularity and loose coupling. This provides flexibility and enables the decoupling of underperforming or redundant service providers.

Enterprise architecture

Within the design principles, enterprise architecture (EA) describes the relationship with enterprise goals, business functions, business processes, roles, organizational structures, business information, software applications and computer systems, and how each of these interoperate to ensure the viability and growth of the organization.

If done well, enterprise architecture (EA) identifies how to achieve separation of concerns and how to define architectural patterns or the rules of how systems communicate with one another. It provides intelligence to the SIAM strategy, bringing a clear definition of the business processes. It plays an important role in the creation and management of a reusable set of architectural domains for the organization. These domains might include business, information, governance, technology, security and others. These structures define the basic principles in which the SIAM controls can be established.

2.7.2.3 Managing process outcomes, not activities

A SIAM model needs to include a delivery structure that supports process inputs and outputs, lays down governance principles, standards and controls, but does not stifle the individual procedural activities that will be proprietary to each service provider in the model. A focus on open standards and interoperability to support workflow, performance management and service management is important. Like any relationship, all parties need to know their role and both receive and provide a valuable contribution. Without this, the relationship lacks purpose or value.

The importance of relationships

In personal relationships, dissatisfaction leads to breakdown and possibly divorce. Similarly, in a SIAM context, an unbalanced relationship may end with the customer terminating a contract with a service provider or the service integrator, or the service integrator or a service provider leaving the ecosystem.

Customer organizations want the benefit of choice and access to specialized service providers that will work with other service providers in the SIAM ecosystem. By focusing on process outcomes, the SIAM model can enable the 'loose coupling' of service providers to allow a relatively easy method to switch them out (and in). Decoupling becomes difficult in environments that have not employed effective strategy and design:

- The transition of service providers needs to happen without affecting the end-to-end service (see section **3.3.3 Transition planning**)

- Knowledge sharing is necessary between incoming and outgoing service providers
- Service providers require an effective exit strategy (see section **3.1.3.10 End of contract**)

In the SIAM model, all service providers understand their role and contribution to process outcomes. They benefit from clear contracts and a relationship that should allow them to innovate and deliver service improvements.

Much of the value of integrated service management solutions is intangible and complex. The value lies in how service elements are defined, provisioned and controlled. Without a SIAM strategy and an approach to service design, the service integrator becomes nothing more than a referee. Without the ability to measure, monitor and decouple, the promised flexibility and choice is not achieved by the customer.

2.7.2.4 IT as a strategic partner

One rationale for adopting a SIAM model is to gain assurance that the IT and business strategies align in relation to the challenges faced and the value sought from a multi-service provider environment. The role and responsibility of the customer organization is critical here. IT leadership can contribute intelligence about technology trends, open market services, opportunities and challenges, but the IT strategy must represent the business or customer correctly.

Having a positive and productive relationship between the business and IT is crucial to effective SIAM. The service integrator acts as the agent of the customer organization and must be seen to be part of it and representing its views (even when externally sourced). Ideally, the service integrator will have representation on the customer organization's management boards, for example, the Executive Steering Board (see section **2.3.7.5 Executive Steering Board**).

The customer organization must define the strategy for services and the model for enterprise architecture (EA), as well as the control exercised through the governance bodies. This may be undertaken in conjunction with the service integrator or trusted advisors, although the customer organization retains accountability.

2.7.2.5 SIAM impartiality

A challenge in a SIAM model is creating an environment where impartiality is both seen and believed to exist. The first imperative is expectation setting. This involves defining the policies and principles governing each service provider. Establishing 'the rules of the game' becomes easier if the service providers have been actively involved in, or at least have a full understanding of, the imperatives. This engagement helps to overcome unrealistic expectations and to confirm a clear understanding of obligations.

2.7.2.6 Clear boundaries of responsibility

Boundaries of responsibility are captured in the SIAM model through service descriptions, service agreements, mandates, charters, processes and operating models. The boundaries of responsibility for each service provider must be well understood to avoid work being duplicated or left undone.

2.7.2.7 Governance model

The creation and communication of a governance model or framework creates parity among service providers. It provides the basis for control and governs activities at all levels in a form to which all service providers can refer (see section **2.3 Establish a SIAM governance framework**).

2.7.2.8 Confidence in the competence of the service integrator

Probably one of the most difficult aspects in demonstrating impartiality and building credibility for the service integrator is through the demonstration of service management expertise and business alignment. When the service integrator is applying process governance and measurement of outcomes, it needs to be consistent and impartial. This will be determined by agreeing how processes are followed and measurements are applied.

2.7.2.9 Trusted relationships

Developing trust and rapport can be a challenge in a multi-sourced environment, particularly if there are legacy service providers and contracts that are not fully aligned with the SIAM model. Establishing strong cross-provider relationships, with effective mediation when challenges occur, is important.

Service providers need to trust the service integrator to act as a mediator and believe the service integrator is interested in helping all parties resolve any issues that might arise. This makes them less likely to question the impartiality of the mediator.

The relationships between the service integrator and all service providers are based on trust, as the service integrator does not own any contracts with the service providers. This does not mean that the service integrator has no authority in the management of the contract, but the service integrator would not typically own the contract management relationship.

2.7.2.10 Service agreements

Service agreements may be contractual commitments, operational level agreements (OLAs) or service level agreements (SLAs), depending on the service provider and customer relationship. The agreements should include explicit service integration targets that focus on service performance, usability and availability from an end-to-end perspective, not just from a service provider's individual commercial perspective.

Service provider agreements should not dictate procedural activities, but should instead exploit service provider knowledge and promote the achievement of end-to-end outcomes and improvements.

Contracts that support integration are important, however, they are not always possible from the outset, owing to legacy or incumbent agreements. SIAM is more effectively embedded into an organization if the service provider contracts reflect the requirement to engage with other service providers and the service integration function. Service providers may refuse to engage with the service integrator if the requirement is not stated explicitly in the contract. This is where 'good' contracts, relationships and governance are vital.

2.7.2.11 Management information

Getting the right data, with the right level of analysis and insight (including data correlation), to the right people at the right time is critical. Effective information management is not easy. There are often many systems to integrate, service providers to manage, tooling challenges, a wide range of business needs to meet, and complex organizational and cultural issues to address. Effective tooling is essential to avoid time-consuming manual effort.

Quality assurance is essential, to help drive efficiency, support the demands of compliance and regulations, and to support decision making. If this CSF is not met, organizations may find that service providers hold on to their own data, ownership is not clear and the end-to-end status is hard to measure and manage.

2.8 Applicable SIAM practices

> **Practice definition**
>
> *"The actual application or use of an idea, belief or method, as opposed to theories relating to it."*[18]

Within the SIAM Foundation BoK, four types of practices are described:

1. People practices
2. Process practices
3. Measurement practices
4. Technology practices

These practice areas address governance, management, integration, assurance and coordination across the layers, and need to be considered when designing, managing the transition to or operating a SIAM model. This section looks at each of these practice areas and provides specific, practical considerations within the Discovery & Strategy stage. Note that the people and process practices are combined and referred to as 'capability'.

2.8.1 Capability (people/process) considerations

During the Discovery & Strategy stage, relevant stakeholders need to be able to assess a wide variety of aspects for suitability, both from a current perspective and a future perspective. These aspects include:

- Business strategy, direction and priorities
- Technology trends
- Current (internal) staff/resources/capabilities
- Existing contracts (duration, pertinence, flexibility, costs, etc.)

This will require varying capabilities and expertise, ranging from 'inside' knowledge of the organization and its strategic direction, to insight into current (and future) technology trends and how they would apply.

In this stage, the outline business case is written, the SIAM strategy (and outline model) created and the project established. Again, a variety of capabilities are required, not least knowledge of SIAM methods and practices, and a 'sales and marketing' proficiency to make sure relevant executives are

---

[18] *www.lexico.com/definition/practice.*

committed to the SIAM strategy/case/project and are willing to allocate resources to the subsequent stages.

Additionally, a governance framework and the policies for roles and responsibilities should be created, which require further capabilities, such as knowledge of legal or contractual obligations, and HR policies and practices (see section **2.3 Establish a SIAM governance framework**).

A broad approach

As much as the Discovery & Strategy stage advocates 'outline', high-level outputs, do not allow it to be the work of one individual (often an enthusiastic IT manager). A narrow perspective at this stage will not only result in a limited output but will also make it harder to gain acceptance from other stakeholders.

A multi-functional, multi-skilled team will enhance both the quality and the acceptance (and thus the success) of the SIAM transition.

Competence is an ability or skill to do or achieve something. If particular skills are needed for the SIAM project, but are not currently present, then common approaches include:

- Buying skills from an external service provider
- Developing skills with existing staff and capabilities
- Borrowing skills

Each of the above options will have implications for time, resource, value of outcome, risk and skill. The challenge in building a SIAM model is that, if the customer organization has immature capabilities, assessing the adequacy of the current state is a challenge (see section **2.4 Define principles and policies for roles and responsibilities**).

The customer organization may reduce the impact of these challenges by undertaking the following approaches (see section **2.2.2 Agile or waterfall?**):

- Pilot/build incrementally, using feedback to refine the approach before building further (Agile)
- Plan, design, build quickly and deploy (waterfall)

Considering incumbent staff displacement

Within both the Discovery & Strategy, and Plan & Build stages, the customer organization must consider the legal impacts of its sourcing choices, specifically regarding incumbent staff. When building the detailed SIAM model, it is important to have a thorough and detailed workforce plan (see section **3.1.6 Detailed roles and responsibilities**).

The wide variety of transfer and other laws in countries around the world can cause challenges if not fully considered. For example, in one part of Australia, there is a government mandate that requires the regional government employer to be mindful of offering opportunities to local businesses and people. The aim is to provide access to contract and job opportunities to the local community as a priority.

In any cross-border business acquisition or disposal, outsourcing, bid to deliver outsourced services or insourcing initiative, it is essential to focus early on how local employee transfer laws will apply in the affected jurisdictions. Obligations may alter the strategic direction regarding the sourcing model.

Not all situations are the same. Considering how to handle the movement and displacement of staff is necessary in situations where:

- A customer organization wishes to bring a service in-house but still assure continuity. It may wish to offer a request to transfer the provider's staff into its organization.
- The customer wants to move a service that has previously been carried out in-house to an external service provider.
- Service providers taking on new contracts may wish to recruit employees from an incumbent provider for ease of ongoing support.
- An external service provider takes on a role previously done by an internal provider, but does not wish to take any of the existing workforce because the work will be relocated, performed in a radically different way or needs significant cost savings.

In some countries, these aims are simple to achieve, but in others are highly complex. There is no global uniformity regarding the application of the legal principles protecting employees (see **Appendix D: Staff displacement legislation**).

Careful approaches to these issues will help to avoid legal irregularity and protect corporate image or brand. Where legal considerations do not exist, this does not imply 'anything goes'. Blatant disregard of any moral and social obligations may lead to former employee dissatisfaction and resulting public stories of poor treatment, leading to potential bad press and reputational damage.

Within the Discovery & Strategy stage, the customer organization should consider the possibilities and limitations of obligations related to retaining staff when the SIAM model is proposed. Input and support from the organization's HR and legal departments, as well as other bodies such as trade unions, should be sought.

The following are considerations for the customer organization, where implications regarding staff are relevant:

- Relevant legal obligations
- Industry participation plans
- Where no legal obligations exist, consideration of moral and social responsibilities
- Lack of expertise in associated legal and HR implications
- Considering role matching between 'as-is' and 'to-be' structures, and where role matches might lead to a requirement to retain an individual
- Planning for staff consultation periods
- Retention, making sure key staff do not leave
- Natural turnover of staff
- Offboarding
- Impacts and constraints of required notice periods
- Costs associated with:
  - Redundancy payments

- o Relocation allowances
- o Cost of outplacement support, retraining and back to work readiness programs for displaced staff
- Staff rights
- Discrimination
- Legal proceedings from unhappy staff

Industry participation plans

Within some local government areas, employers must work within guidelines provided through relevant industry participation plans. These are often created to support growth of a region and its economy, as well as providing more scope for local business and industry development.

In one Australian example, a government plan makes a specific statement regarding people:

"Local businesses are also encouraged to apply best practice approaches in employing, training and retaining indigenous employees and securing joint venture opportunities with indigenous organizations, where available."

2.8.2 Measurement practices

As the SIAM project is not formally active at this stage, there are few targets to measure and report.

From a measurement perspective, focus on the overall commitment of time, costs and resources during this stage, because there is no formally approved business case. Some form of capture and reporting on time and costs needs to be considered, in line with normal organizational practices.

It may be advisable to define the scope of the Discovery & Strategy stage as a separate project, in order to get commitment for the time, costs and resources required to develop the strategy and business case, particularly if it involves a multi-functional team (see section **2.8.1 Capability (people/process) considerations**). Scoping the stage as a project enables close monitoring of time and resource spend against budget.

The most important measurements in this stage are those that will be included in the outline business case. The time and costs of the SIAM project and the quantified expected benefits (including when and how to measure these). These measurements will form the basis of the approval and commitment of the executives and sponsor of the project.

Consider:

- Ensuring the correct understanding to drive decision making
- That decisions are led by data, not just speculation
- Justification and direction metrics for subsequent stages
- Capturing a baseline of end-to-end service levels and user satisfaction for future comparison

Red/amber/green reporting

A useful approach for project status reports is red/amber/green (RAG or traffic light) reporting, where:

- Items going to plan are highlighted in green
- Items at risk of missing planned dates are highlighted in amber
- Items that have missed planned dates are highlighted in red

A further enhancement is to highlight completed items in blue (BRAG).

A high-level SIAM scorecard can be defined with individual key performance indicators (KPIs) aligned to these measurements. The scorecard can also focus on KPIs and metrics aligned to the implementation of the SIAM model, and any other operational KPIs required to achieve those benefits as well as to run operations. The scorecard will evolve and mature as the SIAM project progresses.

2.8.3 Technology practices

Technology and tooling is of limited concern during this stage. The tooling strategy is not defined until the Plan & Build stage, and will address many of the technology considerations. However, building the outline business case and outline SIAM model requires 'an awareness of technology trends', as well as an assessment of current capabilities, which includes technology and tooling.

Technology can have a profound effect on the strategic direction of a business. For example, consider the impact in recent years of cloud-based infrastructure, 'Anything as a Service' (XaaS), or digital disruption.

The outline SIAM model will provide an input to the tooling strategy and, therefore, needs to contain information such as:

- SIAM strategy, including objectives for 'loose coupling' and service provider integration
- Enterprise architecture (EA) assessment

Tools in use during this stage can assist in managing the project or creating an analysis for the outline business case.

CHAPTER 3: SIAM ROADMAP STAGE 2: PLAN & BUILD

The Plan & Build stage is triggered formally on completion of the Discovery & Strategy stage, when the organization confirms its intention to proceed with a SIAM transition. There may be elements that start earlier, to enable stakeholders and the team to prepare for this stage. Depending on the timescales, strategy and approach of the organization, the roadmap might proceed in a linear manner or there may be stages and activities taking place in parallel.

The governance requirements and high-level framework defined in the Discovery & Strategy stage, provide the controlling 'guardrails' for the Plan & Build stage. They ensure the design of the SIAM model is aligned to business requirements and provides all the expected benefits, including the ability to add and remove service providers as needed. The stage objective is to create an adaptable, scalable model that is responsive to the inevitable changes within the business and service provider environments.

The design activities are often delivered in iterations, starting with an initial definition, becoming successively more detailed to complete the design for the SIAM model and create plans for transition during the Implement stage.

3.1 Design the detailed SIAM model

The SIAM model contains many elements, including:

- Service model and sourcing approach
- The proposed SIAM structure
- Process models (including interactions between service providers – see section **3.1.4 Process models**)
- Governance model
- Detailed roles and responsibilities
- Performance management and reporting framework
- Collaboration model
- Tooling strategy
- Ongoing improvement framework

When designing the detailed SIAM model, four distinct architectural viewpoints must be considered:

1. The organization structure for the retained capabilities, any internal service integrator and internal service providers. This should include the formal positions and headcounts as they relate to each other and appear in the HR system illustrating the organizational hierarchy.
2. The process model, showing roles and responsibilities, interactions between parties, ownership and structural elements.
3. The service model, showing service groupings, sourcing strategy and the scope of services allocated to the various service providers.
4. The technology model, showing technologies that will be used to support the other three viewpoints, including toolsets.

All four architectural viewpoints will have their own associated roles and resources that need to be considered, mapped and intertwined. In some cases, these will be different perspectives of the same roles and resources.

Consulting organizations that specialize in SIAM designs can often provide blueprints in the form of SIAM reference architectures. These can expedite the design phase, allowing the customer organization to move more swiftly on to the Implement stage of the SIAM roadmap. However, the suitability of the reference architecture to the customer organization and strategic outcomes must be tested and not assumed. Specialist support can optimize the effort required for the entire SIAM roadmap, particularly the early stages.

3.1.1 Service model considerations

The service model shows the hierarchy of the proposed services, the service provider for each service element, interfaces with services provided by other service providers, and the service assets and interfaces from the customer organization. Additionally, it can show interactions and activities with external agencies that have a governance or legislative influence.

Careful design of this model is critical to success. Iterations will be issued for review and feedback during the design activities. The service model starts with an initial definition and becomes successively more detailed as each iteration is agreed.

3.1.1.1 Service groupings

Adopting a SIAM model will change services and service groupings. It is crucial to first define existing services, their boundaries, groupings and service providers, and assess their suitability for the SIAM model. The design of service groups needs careful consideration, as this will define the scope of services sought from the various service providers.

The service groupings will influence subsequent procurements, with factors including:

- Availability of service providers (internal or external)
- Commercial viability of contracts for external service providers
- Suitability and alignment with common practice
- Ability to substitute service providers if needed later
- Number, and nature, of interactions with other services, service elements and service providers
- Feasibility of (technical) specifications and levels of capability required
- Locality, time zones, languages, local wage levels, economic climate and other geographically specific considerations

If these factors are not considered, it is possible to construct an arrangement that is unattractive to the market. This is a common contributor to the failure of sourcing initiatives, as these will not attract the kind of service provider sought.

Ineffective service groupings

An organization decided to use the following service groupings and to procure a different service provider for each group:

- Networks
- Hosting
- End-user support
- Application development
- Application support

It had a complex legacy application running on an obsolete mainframe that it wanted to continue using. This meant splitting this service up across every group apart from networks.

Although it had several service providers interested in supplying the network services group, there was interest from just one service provider for all the other groups, which included the legacy application.

It can be challenging for the customer organization to change its focus from systems and technology to services as part of the SIAM transition. The existing service provider contracts might focus on technology elements, without a link to the big picture of end-to-end service. The service model design defines the services in scope, the service hierarchy and how the services are to be grouped for sourcing.

Services are categorized, with groups then assigned to specific service providers. The aim is to minimize interactions between different service providers by grouping together similar services to be provided by one service provider. Minimizing interactions makes it easier to adjust the SIAM model if required in the future, and to onboard and offboard service providers efficiently.

The analysis of the current services and service groupings carried out in the Discovery & Strategy stage (see section **2.5.5 Existing services and service groupings**) can highlight which current service groupings are effective, and which are not. This provides useful input to future considerations.

Effective service groupings

A customer organization had outsourced all of its services to two service providers. One was responsible for networks, and the other for hosting, end-user computing and application development.

The second provider performed well in most services, apart from supporting the laptops used by the customer's sales force. Based on this experience, in the design of its SIAM model the organization decided to separate end-user computing into two distinct services – mobile user support and desktop user support – to allow it to procure them in the most effective way.

To derive optimal service groupings, the following approach is suggested:

1. From the Discovery & Strategy stage, take the services that the customer organization wants to consume.
2. Map these to the existing services that are going to remain after transformation.
3. Identify any gaps.
4. From the market analysis undertaken during the previous stage, propose services to fill the gaps. Try to avoid customizing services, instead, using those that are readily available. Focus on services, not service providers, which are considered later.
5. Create a map of proposed services – including services consumed by the business, and underpinning or technical services.

6. Assess the service providers that could provide the services required.
7. Analyze the results and tentatively group services by service provider – avoiding putting too many with one service provider or allocating a service provider to something that they do not normally carry out. Doing so creates the risk that they will be unable to deliver it well or will need to subcontract it, adding cost and complexity.
8. Consider the interactions between service providers – minimize these as much as possible.
9. Revisit and reiterate from point five – until the 'best fit' is obtained.

The output of this activity is the creation of a high-level service model containing all services, grouped by service provider.

One-to-many grouping

It is feasible to have several service providers for one type of 'technical' service. For example:

- There may be two hosting providers to help with service continuity planning.
- There may be several providers of application services, each specializing in one or more applications.

All applications do not need to be allocated to one provider, even if this was the approach taken previously. Although this approach may minimize the number of service providers, it may also drive up cost and reduce service quality.

Most customer organizations are used to dealing with service providers that deliver systems, email, internet access, etc. A traditional method of managing these service providers entails specifying technical requirements, with the customer organization taking an interest in the build and playing a design authority role, either as lead or as a stakeholder.

Moving to a SIAM model requires the customer organization to focus on outcome-based services, for example, a cloud-based solution that fits its requirements but requires little technical or design input from it. This can cause some initial anxiety as the direct control the customer organization is used to is no longer needed (or recommended). Moving to a SIAM model can require the customer organization to develop different capabilities, focusing on governance and supplier management, instead of technical detail.

Eyes on, hands off

Customers often try to retain a level of direct control, unless they have undergone an organizational change management (OCM) process that provides clarity and confidence in the planned changes. The service integrator will need to work hard to reinforce appropriate behaviors through all of the SIAM layers.

Moving to a SIAM model is not simply a move to a multi-service provider environment (which many customer organizations already have), but is a move to integrate and control that environment to better meet business needs. This is the challenge that the service integration layer is expressly created to manage. The key to managing services, as opposed to systems, is to ensure that all parties have a clear and documented understanding of the requirements and goals and where they fit into the end-to-end service.

This is where service level agreements (SLAs), critical success factors (CSFs) and key performance indicators (KPIs) become crucial in the definition of requirements and commitments to deliver them. The links between these documents are also crucial, ensuring that the end-to-end service is understood. **Figure 11** shows the relationships between these elements.

Figure 11: Relationships between targets and performance documents

SLA flow down

An 'SLA flow down' can be helpful when piecing together all the SLAs, OLAs and underpinning contracts in a service. It shows where any anomalies lie and can be the basis to renegotiate the agreements.

A customer organization no longer needs to act as design authority for the resilience of technical components or provide lifecycle management of the infrastructure. Instead, it needs to define required levels of availability, security, performance and utilization to the service providers that will then ensure these measures are met. The contracts (or agreement with an internal service provider) will outline the consequences of failure, such as service credits.

3.1.1.2 Developing the SIAM model

The specific form in which the SIAM model is defined will vary from organization to organization. The format itself is not important. What is important is that it is in a form that enables all stakeholders to understand fully:

- The SIAM model as a whole
- Their scope and context within it

The model should be both holistic and detailed, clearly defining all service elements, touch points, roles and responsibilities. At a high level, a simple service chain diagram (see **figure 12**) gives an example of a 'helicopter' view, which provides the context for each of the individual service provider's services and the primary engagement paths between them.

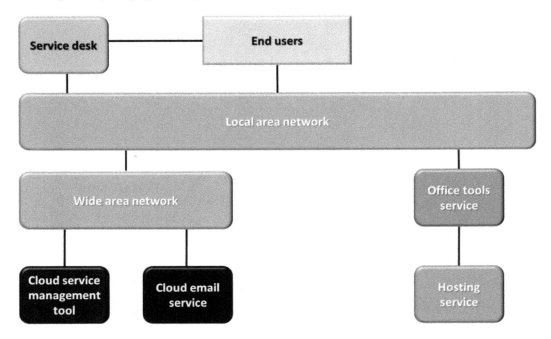

Figure 12: Service chain diagram

At a more detailed level, a range of techniques can be used to help ensure that the SIAM model is fully defined and understood.

OBASHI® dataflow analysis

One technique to help service mapping is OBASHI. In **Figure 13,** a dataflow analysis is provided as an example. The elements of the Business and IT (BIT) diagram that link to support the business process are clearly shown, and the relationships between them can be easily understood. Arrows are used to show which way the data is flowing.

The start and end points must be clear so that the end-to-end dataflow can be understood.

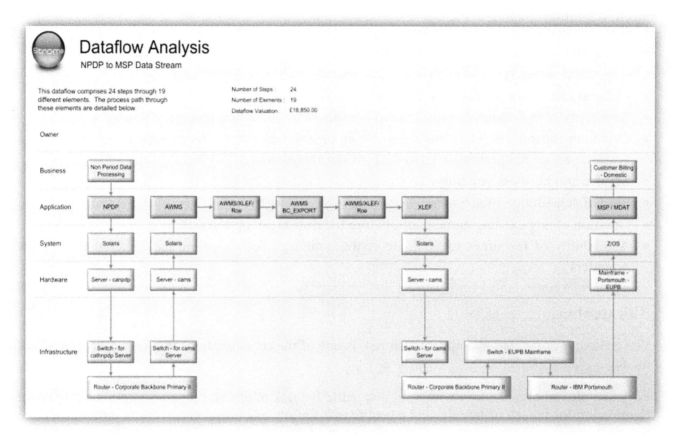

Figure 13: Dataflow Analysis View (DAV)
Reproduced under licence from OBASHI Ltd

As well as a visual representation, the SIAM model should detail how each component interacts with others. The model should include a description of each component and information on how it fits into the framework. Additionally, there should be accompanying information describing the governance mechanisms in detail using the service model diagram as a reference. Diagrams explaining escalation mechanisms, escalation flows and escalation contact information are required.

Different audiences will require different data from the SIAM model, so it is important for each stakeholder to be able to view simple or complex data, depending on their requirements.

The business context must also be clear and related to the SIAM model, for example:

- Is the business heavily regulated?
- Are there governance requirements or legislation to consider?

3.1.1.3 SIAM transition approach considerations

There are some important considerations of the likely transition approach that can affect the design of the detailed SIAM model. These should be considered initially as part of the SIAM strategy within the context of the four stages of the SIAM roadmap (see section **2.6 Define the strategy**).

It is important to consider the outline implementation plan for the SIAM model at this stage. The plan should consider the situation and requirements of the customer, as well as those of the other stakeholders (see section **3.3.3 Transition planning**).

The considerations include:

- The customer organization's appetite for risk
- Incumbent service providers' presence, power and willingness to change
- Cultural factors for all stakeholders, such as attitude to change
- Sensitivity of the customer organization's business to disruption to service levels
- Customer capabilities, including commercial, operational, service management, business change, project, service integration and service provider management
- Service and business volatility
- Ease of substitution of services
- Duration of any existing agreements, before the next break option arises
- Availability of resources to facilitate change, manage collaboration and onboard new service providers

---

Risk appetite

The customer organization's appetite for risk is one of the considerations that can affect the SIAM model.

For example, an organization with a high appetite for risk might be prepared to pursue a contract termination for breach of targets, and a hostile exit for any existing contracts with external service providers that will not be part of the SIAM model. This would also depend on a customer organization with a high tolerance for potential disruption to service levels.

Each of the considerations focuses on areas of risk to the business. Managing the transition from one service model to another, or from one service provider to another, involves risks to service performance and continuity. The customer organization needs to ask:

- What is our overall approach to the implementation of the change (where do we start, where next)?
- How fast are we able to move?
- Which stakeholders do we engage first, second, third … ?
- Which services do we involve and when?
- What are the milestones in delivery?
- How do we work with the outgoing incumbents during the period of transition?
- What resources are available?
- What resources are required?
- How can any gaps in resource be bridged?

Initial planning iterations are useful to eliminate unfeasible approaches for migrating to the desired SIAM model. The final plan is likely to combine elements from several scenarios. The customer organization should not be too quick to discount options as improbable, they may have good elements or help the team to think laterally.

3.1.2 Sourcing approach and the selected SIAM structure

All SIAM models will include a variety of service providers, some that are newly appointed and some with renegotiated contacts.

A successful SIAM model should provide the ability to manage change over time. There is likely to be a mix of sourcing options in the environment and these arrangements will change over time based on the evolving needs of the customer organization. This has many implications that need to be considered as part of the SIAM model, for example, in the ownership or licensing of intellectual property (IP), tooling and exit obligations.

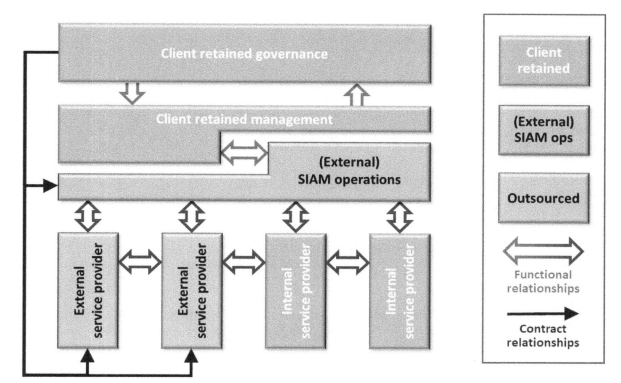

Figure 14: SIAM ecosystem 'draft'

SIAM models vary considerably, but they will share common attributes, including a mix of service providers, both internal and external. The operational/functional relationships (represented by the blue and green arrows in **figure 14**) should be supported in a robust SIAM model and include:

- Peer-to-peer relationships between service providers
- Service provider to service integrator
- Customer organization governance and management to service integrator

The contractual relationships (represented by the red arrows in **Figure 14**) include:

- Contracts with the various external service providers
- A contract or agreement with the service integrator (depending on whether it is internally or externally sourced)

In a SIAM model, the customer owns the contracts with all external providers. There is no concept of a 'prime' vendor, which has a contract with the customer and contracts directly with and manages other providers. Some providers may still employ subcontractors, but the provider remains fully responsible for their services and all obligations in their contract. Many contracts include clauses that prevent subcontracting without approval from the customer.

Resource augmentation

Where resource augmentation is used to supplement an internal service integrator capability, this should not be considered as an external service integrator. Instead, the contracts for resource augmentation are traditional staff supply contracts.

Resource augmentation contracts in a SIAM model should include separation of duties to avoid potential conflicts of interest.

The purposes of contracts are twofold in a SIAM ecosystem:

1. Establish the contractual relationship and certainty that any traditional externally sourced contract should provide
2. Support the SIAM strategy through the relationships shown in figure 14

3.1.2.1 Sourcing the service integrator

The decision about how to source the service integrator is an important one.

The financial benefits of a service integrator may be difficult to measure, as integration is mostly non-transactional and deals with issues avoided as much as benefits realized (see section **1.6 SIAM structures**).

Some organizations may not see the need to seek an external service integrator, and instead source the role internally. Where a customer organization has the right capabilities, this can be a good approach. If the customer organization does not have the right capabilities, it needs to create them. For some, this challenge is modest and quickly achievable. For others, it is much more difficult.

Because an internal service integrator does not have a legal contract with the customer organization, it can be more flexible and accommodating to change than an external service integrator. However, an internal service integrator will still potentially require extra budget and resources if its role changes significantly.

If a hybrid structure is adopted, the 'split' between the retained and externally sourced roles in the integration layer needs to be considered. This might be established as a hierarchical structure (customer staff over external service integrator staff) or a vertical one (side by side).

3.1.2.2 Planning the role of the service desk

The sourcing of the service desk will vary from SIAM model to SIAM model. The organization providing the service desk should be considered a service provider in the SIAM model, whether it is provided by the customer organization, the external service integrator or one or more service providers. It should be managed in the same way as all other service providers.

Although a service desk can be seen as just another service to be provided, it is worth considering in detail, owing to the role it plays in contributing to customer satisfaction, information sharing, support for integration activities and managing issues across SIAM layers and stakeholders.

The service desk is often seen as a good candidate for external sourcing because of factors including high staff turnover and high management overhead. Some organizations, however, prefer to source the service desk internally or use a hybrid approach. The customer organization must decide whether it sources the service desk internally or externally.

The approaches discussed in this section are:

- Service desk provision by the external service integrator
- Service desk provision by the customer organization
- Service desk provision by an external service provider
- Multiple service providers providing separate service desks

---

Case study

AIR is the second largest South African provider of mobile technology and fixed telephony, and is also a service provider of broadband and subscription television services. The brand is operated under the AIR brand name, however, internally relies on many other subservice providers to provide specialized services:

- Service provider A provides subscription television services
- Service provider B provides broadband services
- Service provider C provides network equipment for mobile technology
- Service provider D provides applications to manage mobile technology
- Service provider E provides cloud services as part of continuous delivery

AIR began with a centralized service desk and, after 18 months' operation, was facing issues where customer complaints indicated it took a long time to receive a response to standard queries and requests.

AIR ICT wanted to empower the service desk with the ability to resolve more requests at first contact. However, the service desk struggled with the complexity of knowledge required across mobile technology, fixed telephony, broadband and subscription television services. The customer satisfaction for the service desk, especially on subscription television and broadband services, was poor.

AIR ICT hired an external consultant to help it to consider how to improve the service desk. One of the recommendations was to reconsider the service desk sourcing strategy. A move to a service desk sourced through an independent best of breed service provider, utilizing technology to route calls to specialists of each service, improved customer satisfaction immeasurably.

Within a SIAM model, the service desk acts as a 'single source of truth' for consumer satisfaction and provides important management information about service performance. If the service integrator is not providing the service desk, it must work very closely with it and use the service data it provides.

The service desk plays an important role in the day-to-day contact with the consumers of the service, providing crucial metrics around satisfaction, both quantitative and qualitative, for the service integrator. There is no best option for the service desk, but there are a range of advantages and disadvantages to consider.

Service desk provision by the customer organization

In this option, the customer organization retains the provision of the service desk function, usually alongside an internal service integrator. The rationale for the customer organization providing the service desk is usually related to a desire for control and the benefit of internal business knowledge, shared values and culture, or where it already has an established, mature service desk capability.

There may, however, be specific reasons for the customer organization to retain the service desk in-house, such as legal or regulatory requirements in, for example, the defense industry, where sensitive information is not allowed to leave the customer organization.

Advantages

- The ability for the customer organization to control/influence the service quality and delivery from the service desk directly.
- No confusion around ownership of intellectual property (IP).
- A perception from end users of ownership and end-user orientation.
- It may be possible to provide a 'local touch' as service desk staff can be located in customer organization offices, rather than in an offshore or service provider location.
- The business process knowledge of service desk staff may increase competence and allow better evaluation of the severity and urgency of issues.
- Ability to address a wider scope of queries, based on a wider knowledge of the business.
- No complex contractual structures.
- The potential for a higher degree of transparency about service performance, as there is no incentive for the service desk to hide poor performance figures.
- The internal service integrator and service desk being part of the same organization helps to standardize processes and toolsets, for example, for incidents or request records, enabling a 'single source of truth'.
- Efficient collaboration and cultural orientation between the service desk and internal service integrator functions.

Disadvantages

- Quality outcomes depend, to a high degree, on the process and tool maturity within the customer organization, rather than those of a specialized service desk provider.
- It may be difficult to find appropriately skilled resources within the local labor market.
- Career path and growth opportunities for service desk team members must be considered and managed.
- Risk of the service desk operating in a silo and not building relationships with service providers.
- Staff optimization based on volume trends (upscaling/downscaling) will be more difficult with internal resources. An external service desk provider may be able to scale more easily.

Case study

GBP University of Business Studies is the seventh oldest university in Australia. It has an annual revenue of A$120 million. Its faculties and schools include:

- Arts, Business, Laws and Education
- Science
- Engineering and Mathematical Science
- Health and Medical Sciences
- Business and Technology Management

GBP University has a centralized IT support desk that supports two campuses. The support desk manages the tickets and complaints logged by users of the university. The support desk had no mechanism to classify incident tickets. The support desk technicians were in a constant firefighting mode to handle tickets and workload.

An external business analyst conducted a gap analysis and documented the following challenges with the current model of centralized IT support desk:

- Lack of knowledge of best practices associated with service desk and service management processes
- No defined KPIs for centralized support desk
- No focus on automating repetitive tasks
- No concept of a service request

As GBP University wanted to improve its customer experience, the first step was to rethink how to improve the maturity of the current centralized IT support model and the team, as it had no in-house service desk capabilities. This might be an ideal candidate for external sourcing from a specialized service desk provider.

Service desk provision by the external service integrator

In this option, the external service integrator provides the service desk. The service integrator acts as a lead supplier, as through the service desk it also provides a service.

This structure can provide a good alignment between the service desk and the service integrator, and will typically be chosen when the organization acting as the external service integrator has additional and mature capability to provide the service desk. Where the external service integrator directly provides the service desk, the following advantages and disadvantages should be considered.

Advantages

- The ability for the service integrator to influence the service desk outcomes, which may positively influence the end-user experience.
- The service integrator and service desk being part of the same organization helps to standardize processes and toolsets, for example, for incidents or request records, enabling a 'single source of truth'.

- Efficient collaboration and cultural orientation between the service desk and service integrator functions.
- The combined role may be more commercially attractive for an external service provider.

Disadvantages

- The service desk needs to be considered as another provider in the SIAM model. Integrating the service desk with the service integrator layer may compromise – or be perceived to compromise – impartiality in managing service desk performance.
- If the service desk performs badly, this could affect the reputation of the service integrator and compromise its ability to perform effectively and build relationships.
- The combination of the service integrator and the service desk, especially when captured in a single contract, can provide future challenges when intending to change one without the other.

Service desk provision by an external service provider

The service desk is considered a service like any other as part of a SIAM model. Service providers in this space may bring specific expertise, tooling, flexibility, agility and scaling that would otherwise be difficult to achieve.

Advantages

- Enables the customer organization to focus on its strategic direction and business objectives rather than day-to-day user based transactional and operations management.
- The service integrator and customer organization are separate from the service desk function and, therefore, can look at service provision from an independent perspective.
- Sourcing from a specialist service provider will often provide expertise and enhanced performance.
- A more economical contract may be negotiated based on the actual volumes and can incorporate penalties/discounts, etc.
- The service desk will be treated as any other service provider and form part of the collaborative, innovative culture.

Disadvantages

- There is no guarantee that the external service provider can assimilate business knowledge effectively, it will take time.
- The service may experience stages of good or bad performance when the service provider's staff change.
- Service desk resources might be shared across multiple customers and some customers might be given higher priority.
- Security and intellectual property (IP) considerations need to be addressed.

Case study

TRE Company Ltd is the largest pharmaceutical company in Japan, and a top 10 pharmaceutical company in the world. The company has more than 15,000 employees worldwide and achieved ¥10 billion in revenue during the 2015 financial year.

The Information System (IS) department of TRE is a multi-service provider environment:

- **Application delivery:** TRE internal
- **Infrastructure operations:** service provider A
- **Service desk:** service provider B
- **Business intelligence and reporting:** service provider C

Driven by the need to manage the results for the typical multi-service provider ecosystem, TRE information systems (IS) decided to form a service integrator function. Based on the market capabilities and previous experience, service provider B was contracted for the service integrator role.

Service provider B appointed a new service manager to manage the integration layer, in addition to the existing service desk manager to maintain the independence of operations between two functions. The internal organization structure of service provider B indicated that the service manager was two levels higher in the hierarchy than the service desk manager. The service manager managed all costing, budgeting and resourcing decisions of the service provider B on the TRE account. Most of the service desk plans needed an endorsement from the service manager before they could be published or discussed with other service providers and the customer organization.

It is important to ensure that the service integrator is at the relevant organizational level and has the appropriate autonomy if the duties of the integrator role are to be achieved successfully.

Multiple service providers providing separate service desks

In this option, different service providers provide their own service desks and toolsets, and the service integrator provides a consolidated view. This is only an effective choice where it is clear to the consumers of the services which service desk to contact for support. For example, the payroll department can contact the service desk of the financial application provider directly.

Note that, often, different service providers have their own existing service desk. These service provider service desks can provide a second-level escalation point for the service desk supporting the users of the services within the SIAM model. Allowing the SIAM model's consumers to access multiple service desks removes the single point of contact and can create confusion.

Advantages

- Improved knowledge management, as these service desks are within the service provider that delivers, manages or supports the specific service.
- Possibilities for better and faster support, as each service desk's scope is limited, driving specialization in skillset.

- Better ability to respond to volume spikes or drifts, as they will only affect one service desk, not all service desks, if related to an issue with a particular service.
- Poor performance can be addressed very specifically, as it will only apply to one service desk.

Disadvantages

- Multiple service desks will increase the risks of different processes and toolsets being used for similar activities. The service integrator must facilitate end-to-end transparency across all the service desk providers.
- There can be higher reporting overheads because the data to be processed will have to be collated from several service desks.
- Reporting may be difficult or impossible if the various service desks use and report on irreconcilable metrics.
- Risk of inconsistent end-user experience (if one service desk is providing an excellent service while others may fall short …).
- Potential for inefficiencies when consumers contact the wrong service desk, or 'hop' between service desks to find a solution. This will require cross-service provider escalations and, possibly, service integrator coordination.
- Confusion and/or frustration in users struggling to know which service desk to contact.
- Risk of 'blame culture', where different service desks 'bounce' incidents or requests to one another.
- Complexities and overheads involved in evaluating and processing customer satisfaction surveys across the multiple service desks.

3.1.3 The importance of contracts in SIAM

Creating contracts that are robust, usable and appropriate for use in a SIAM environment is critical. Contracts support an environment where there is a shared understanding of requirements, and clear accountabilities and responsibilities for both parties to the contract – the customer and the service provider (see section **2.3.13 Supplier and contract management**).

SIAM requires contract structures that differ from non-SIAM agreements, which focus solely on the responsibilities of a single service provider. Contracts in SIAM models should try to link service provision across the full ecosystem, encouraging acceptance of, and adherence to, a common set of delivery rules and governance, with the common purpose of supporting the customer organization's services and desired outcomes.

As well as outlining the services and service levels to be delivered by individual providers, including any external service integrator, the contracts must address integration. This includes areas such as tooling, process integration, knowledge management, collaboration and participation in the structural elements. Integration is the key distinction between a more traditional contract and a contract appropriate for a SIAM environment.

Addressing integration in contracts underpins and reinforces the 'one team' approach necessary for success in SIAM. It also provides a strong foundation for the service integrator, facilitating collaborative behaviors and helping to avoid a blame culture from the outset. Contracts that are designed for use in a SIAM ecosystem provide a necessary focus on end-to-end service delivery, regardless of which service providers are involved in service provision.

Look before you leap

It is often the case with a SIAM implementation that the early roadmap stages of Discovery & Strategy, and Plan & Build consume much of the available resources in getting SIAM contracts right. This can take more time than the actual implementation activities.

This is important for all parties. Underestimating the scale of the Discovery & Strategy, and Plan & Build activities, and the overhead on each of the SIAM layers even before the ecosystem is up and running, can have devastating effects on the quality of the implemented SIAM model and the services it delivers.

The customer organization needs to provide enough detail around acceptable ways of working, collaboration and integration activities within the contracts, while still allowing for individual service provider approaches. Constraining service provider activities can lead to the customer organization not benefiting from specific provider specialisms and creating contracts that inhibit the ability for service providers to perform effectively and efficiently.

Less established or less mature service providers may not have an established service portfolio or corresponding service delivery capabilities. This has the potential to influence the overall ecosystem capabilities negatively. Consequently, however, they may adapt to the required ways of working more easily.

There are also risks associated with mature service providers. When established service providers are selected, it is often with a high emphasis on promised cost savings. The cost savings are typically based on exploiting economies of scale and offshoring models, with staff being sourced in lower-cost countries. There is a risk that these staff are highly trained on the service provider's existing processes, tools and established, specific ways of working, but not on the specifics required for the future customer's requirements, desired outcomes and chosen SIAM model.

Education, skills and experience can constrain a potential service provider's ability to adapt to a SIAM model and ways of working without compromising its own benefits through economies of scale. This may result in it not bidding for the work or taking the work without fully understanding the implications for itself. It is important for the customer to assess potential service provider's capabilities before awarding a contract.

Some customer organizations engage external consultants to assist with contract design. These consultants must focus on value, not just simplification and cost optimization.

If appointed early in the SIAM roadmap, the service integrator can work with the customer organization to review existing service provider contracts, identifying duplications, gaps and any inconsistencies with the future SIAM model. This information can then be used to design new contracts that are tailored for the customer organization's particular SIAM model.

The contracts must contain the right level of detail. It is important to review contracts against the SIAM model before sending them to potential providers. The nature of the roles in a SIAM ecosystem require close teamwork between contract experts working on the design and service integration experts, who are often focused on implementation. Contracts must not constrain, but enable. This is possible only by using the right expertise at the strategic and planning stages.

Investing in expertise at the enterprise architecture (EA) and solution architecture levels early within the Discovery & Strategy stage allows an effective foundation to be built for the SIAM model, including appropriate contracts.

If a service provider could also be acting as the service integrator, then impartiality is essential. Separation of concerns is an important principle, as is working openly with all chosen providers in establishing the overall solution.

Many contracts in traditional sourcing environments are designed to discourage future changes to responsibilities and activities because of the expected overhead of contractual, legal and commercial considerations. SIAM contracts need to allow for regular alignment and realignment to support continuous improvement within the SIAM model. Significant change in one contract, such as the introduction of a new service provider, should be reflected in the review of neighboring contracts.

3.1.3.1 Contract arrangements

In a SIAM environment, the customer should have signed contracts that align with the SIAM model in place with all external providers. Although this might seem obvious, it is not uncommon for organizations, even of a global size, to be using service providers with whom they have no formal agreement or where a formal agreement has expired.

Often, this happens away from head office, typically in geographic areas where the organization does not have a large presence with all the supporting infrastructure and functions, so staff resort to using local services. This may begin on an infrequent, informal basis but gradually dependency grows, with the lack of any formal agreement only being highlighted when service is interrupted, and the organization is adversely affected.

Contracts need to cover responsibilities throughout the whole contract lifecycle. All too often, contracts concentrate on the initial set up and operation of the service but contain very little about ongoing improvement or what happens at the end of the contract, whether that is the transfer of services to another provider or termination of the service. This can result in handover issues, disruption, delays, loss of knowledge, loss of intellectual property (IP) and, potentially, the loss of data.

Operating without a contract

A new start-up venture providing legal services and staffed largely by lawyers decided that it needed the services of an applications developer to build an ecommerce site. One of the non-executive directors had seen a developer at work in another company and suggested he would be suitable. The developer was asked to present and was engaged. No contract was signed.

After two years' work and the expenditure of many thousands of dollars on development, little had been achieved. The newly hired operations director of the start-up expressed no confidence in the developer and decided to remove him. The developer claimed he had delivered what had been asked for, but that requirements continually changed.

An investigation into the competence of the developer showed grounds for the customer to sue. To avoid the cost, effort and the potential for reputational damage, the parties settled out of court and parted company. A formal contract would have provided a simpler and more cost-effective way to terminate the arrangement.

Within a contract, there should be a clear indication of how the contract will be managed, including:

- Escalation paths
- Procedures for agreeing contractual changes
- Extensions
- Handling disputes
- Exit plan

It is not good practice to rely on individual contract templates, of varying quality, from a wide range of different service providers, as these will often be geared towards protecting the service provider rather than the service consumer. Organizations should look to develop their own standard contract templates that can be tailored to the specific services required, but which contain standard governance controls, lifecycle considerations and terms and conditions, all thought out in advance and checked by the organization's legal representatives.

Model services contract

An example is the UK government's standard contract templates and in particular the Model Services Contract (MSC) (available at the time of publication here: *www.gov.uk/government/publications/model-services-contract*).

"The Cabinet Office and the Government Legal Department have published an updated version of the Model Services Contract. This version reflects developments in government policy, regulation and the market. The Model Services Contract forms a set of model terms and conditions for complex services contracts that are published for use by government departments and many other public sector organisations ...

The documents are intended for use by procurement specialists and lawyers. You should carefully assess whether the provisions contained are appropriate for your project requirements and tailor the documents appropriately if required. Any specific and/or specialist legal advice should be sought from your legal advisors.

The Model Services Contract:

- *has been developed for services contracts with an average, annual value over £20 million*
- *has been developed for complex and high-risk contracts*
- *reflects current government priorities and recommended ways of doing business at the time of publication*
- *aims to aid assurance and reduce administration, legal costs and negotiation time*

> *It is suitable for use with the range of business services that government purchases and contains applicable provisions for contracts for business process outsourcing and/or IT delivery services."*[19]

The service integrator needs to be recognized as the customer's agent, and this needs to be reflected in both service integrator (where externally sourced) and service provider contracts, using 'right to manage' clauses to give the service integrator a legal basis for its activities. Incorporating standard, aligned service performance measures in the contracts will also assist in the end-to-end performance management and reporting.

Some service providers, for example those offering commoditized cloud services, may mandate the use of their own standard contracts. If this is the case, it can still be a significant advantage for the customer organization and the service integrator if a large proportion of the contracts in place are in a relatively standardized format and content, tailored to support the SIAM model.

Within the contracts, some of the schedules in use may vary by type of service. For example, service levels for hosting are likely to differ from application development and support. There are often differences in contracts between certain types of provider – strategic, tactical, operational, commodity. What is sensible for a large strategic provider may be unnecessary for a small operational provider.

Other contractual considerations in a SIAM environment include:

- Contract length
- Contract change triggers
- Alignment of contract start/end dates. In a perfect world, all contract start and end dates in a SIAM model would align, but this is not usually possible, as it would also cause significant overhead and risk to replace all contracted service providers at once (see section **4.1.1 Big bang approach**)
- Dependencies with other service providers' contracts or internal agreements
- Data ownership and access
- Links to supporting schedules and enabling documents
- Offboarding and cancellation charges

In SIAM, contracts are often structured in order to minimize their differences. Requirements that are the same across a number of service providers can be structured into schedules. These can then be reused in individual contracts without change.

Although there will always be unique elements in contracts between the customer organization and each external service provider, many SIAM implementations use a generic contract structure that maximizes reuse of common schedules. Using this approach can:

- Speed up contract drafting, as common requirements can be drafted once and reused, and similar requirements can be copied from another contract and then edited if necessary
- Allow more time to be focused on understanding and drafting any unique requirements for a specific contract

---

[19] *www.gov.uk/government/publications/model-services-contract*.

- Facilitate comparison between contracts
- Improve understanding of the contracts
- Facilitate the addition and removal of specific services
- Make subsequent contract changes easier to achieve, by limiting the changes to specific parts of the structure

The design and application of a generic contract structure will depend on the specifics of each SIAM model.

Table 3 shows example options for the design and application of a generic contract structure, and the advantages and disadvantages of each.

Table 3: Advantages and Disadvantages of Contract Structures

Contract structure	Advantages	Disadvantages
Same structure for all contracts	• Maximizes reuse between contracts	• Can be a challenge for some service providers (for example, commodity service providers) to adopt the same structure • Requires careful design to avoid a situation where certain contract schedules are not relevant for particular service providers (for example, some service delivery requirements are not applicable for the service integrator)
Unique structure for the service integrator contract; same structure for all service provider contracts	• Removes the need to include service delivery related schedules in the service integrator contract • Maximizes reuse in service provider contracts	• Can be a challenge for some service providers (for example, commodity service providers) to adopt the same structure
Unique structure for the service integrator contract; more than one structure for the service provider contracts	• Removes the need to include service delivery related schedules in the service integrator contract • Allows structures to be tailored to the needs of different types of service provider (for example,	• Requires careful design to maximize reuse and avoid duplication between structures • Alignment between service providers is challenging

	commodity service providers)	
Different structure for every contract	• Allows all service providers to have a structure tailored to their unique needs	• Limited ability to reuse schedules • Difficult to compare contracts • Extends drafting time

Typically, contract structures for SIAM use these components:

- A master services agreement (MSA), sometimes referred to as a 'head agreement'. This defines the overall contractual relationship between the customer organization and the service provider. The MSA can be common to the structure of all contracts in the SIAM ecosystem, or only used in the structure for the service integrator contract, with a reference to it in the service provider contracts.
- A 'common services schedule', which includes all of the requirements that are the same for all service providers. This schedule is then reused without change in each of the individual service provider contracts. Requirements of this type can include:
  o Intellectual property (IP) rights
  o Non-disclosure requirements
  o Force-majeure statements
  o Liabilities and waivers
  o Rights to assign or novate contracts
  o Contract change and termination arrangements
  o Payment terms
  o Contractual dispute process
- A number of supporting schedules, each including the requirements for a particular topic. Each schedule defines requirements that are not subject to regular change and is the same for all services provided by that service provider. Supporting schedules can also be reused across multiple service providers. Requirements in supporting schedules should not also be in the MSA or anywhere else in the contract. Examples of supporting schedules include:
  o The approach to be taken for testing and deploying new and changed services
  o Detailed security requirements
  o The performance management approach
  o Service management requirements
  o Disaster recovery requirements
  o Service desk requirements (where appropriate)
  o Technology refresh requirements
  o Standards that should be followed

Although it may seem logical to include some of these schedules in an MSA or common services schedule so that they are applicable to all suppliers, putting them into supporting schedules allows variation to meet the needs of particular service providers and geographies.

- One or more statements of work (SoWs). These contain requirements that are specific to the services provided, for example service levels, charges and hours of service. There can be a single SoW that includes all services from that service provider or a separate SoW for each service.
- A common document that describes the way all parties should interact, for example a collaboration agreement. This is sometimes included as a contractual appendix.

Figure 15 shows an example contract structure that has a unique structure for the service integrator, including an MSA, and a common structure for the two service providers, including a common services schedule (although service provider 2 has multiple SoWs).

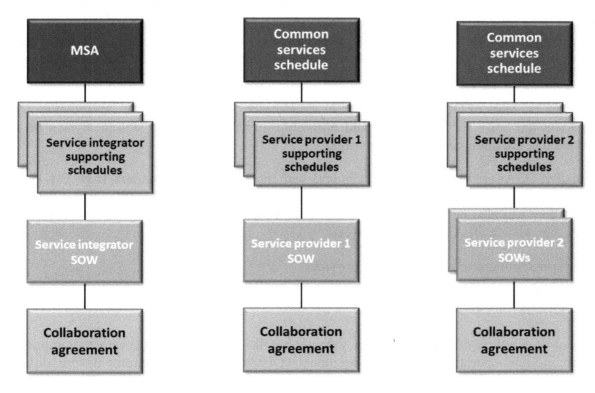

Figure 15: Example of a generic contract structure

It is important to ensure that contracts are unambiguous and specific. If this is not the case, there is a high risk that unnecessary effort will be spent after implementation debating specific points to avoid service credits or additional costs (see section **3.1.3.9 Dispute management**). One means of avoiding or mitigating such a risk is for the contracts to include definitions of specific terms, sometimes known as standard contract terms (SCTs).

Ambiguous contractual requirements

A contract included a requirement for 'evidential testing' to be carried out for every application release. The meaning of this was debated between the customer and the service provider, which held up milestone payments for delivery.

It took some time for the customer to agree that it also did not know what was meant. The phrase had been copied from a boilerplate contract previously used for security systems and applied to the use of CCTV cameras recording evidence.

A contract might not necessarily include details related to supporting items (for example, process definitions), instead including a reference to these. This can provide flexibility when maintaining the referenced items without having to renegotiate the contract. Note that these supporting items are still subject to change control from a contractual perspective.

3.1.3.2 Non-contractual arrangements

In SIAM models, contracts can be supplemented by non-contractual 'ways of working' documents. These are often referred to as operational level agreements OLAs and include guidelines and common ways of working that support effective integration and delivery across the ecosystem.

OLAs tend to be specific to a particular process (for example, incident management) or practice (such as tooling), and are often developed by the appropriate process forum or working group. The use of OLAs can underpin any formal collaboration requirements and help to create a culture where service providers work fairly and honestly with each other and with the service integrator.

Although the contents of OLAs are not themselves contractual obligations, it is important to define in the contracts the basis for developing and enforcing any OLAs. OLAs should always be formal agreements that are documented and controlled.

Traditionally, OLAs are only used for agreements with internal service providers. In SIAM, OLAs can also be used with external service providers. Although some OLAs are often common across all service providers, they can also be used with a smaller subset to define how two or more service providers work together. Hence OLAs can exist between:

- The service integrator and all service providers
- The service integrator and specific service providers
- Individual service providers

OLAs must be carefully designed so that they do not conflict or overlap with contractual requirements. They can include operating level measurements (OLMs). These measurements are defined by deconstructing service level commitments, deliverables and interactions that involve more than one party and set targets for each.

OLAs can be brought together into a single operational level framework (OLF), which is managed and controlled by the service integrator and shared with all service providers. The OLF can also contain other information that describes how the SIAM model operates in practice, including a summary of service levels and common requirements. This can help to address issues where competing service providers are unwilling to share the full details of their contracts with other service providers.

The service integrator is accountable for coordinating the overall development, communication and management of OLAs, OLMs and the OLF.

SLAs, OLAs and OLMs

An SLA contained a target for all service providers to restore service in 4 **hours or less**. A working group with representatives from the service integrator and the service providers broke down this target by assessing the contribution made by each party to the end-to-end incident management process.

They agreed an OLA containing lower-level OLMs:

- The first-level service desk had **15 minutes** to investigate before routing to the appropriate second-level service provider resolution group
- Each service provider had to escalate unresolved incidents to its third-level support **within 1½ hours**

This helped to identify issues with particular interactions so that they could be addressed. As a bonus, the end-to-end support model was now fully mapped and understood, enabling the incident management process forum to carry out regular checks of the end-to-end process, identifying bottlenecks and opportunities for improvement.

OLAs can be used as a way to test that changes to ways of working achieve positive outcomes before committing to contract amendments. This can avoid the expense of making contract changes that subsequently do not provide any benefit after implementation.

Using OLAs to test new ways of working

Several months after going live, the service providers identified that their contracts did not allow them to 'stop the clock' for an incident when they were unable to contact the affected user for more information.

The incident management forum decided to design a suitable approach, which was developed, agreed and documented in an OLA.

All service providers trialed this new approach and, once the forum members agreed, the service integrator asked the customer to raise a contract change with all service providers, containing the approach defined in the OLA. All accepted it at zero cost.

OLAs, OLMs and any associated operational level framework should be defined or changed when the need is identified, ideally through workshops attended by all parties. This can be during the Plan & Build stage, or it may be during the Implementation, or Run & Improve stages.

3.1.3.3 Collaboration and contracts

Irrespective of any collaboration agreements, contracts in a SIAM ecosystem should include requirements for collaborative working. The aspects that need to be considered can be split into three categories:

1. Contractual
   The contractual category includes explicit requirements that are mandated in a contract:
   o Principles for collaboration and cooperation
   o Role of the service integrator
   o Assurance responsibilities
   o Delegated authorities
   o Right to manage/managing agent
   o Approach to end-to-end service levels

    o Definition of critical deliverables
    o How culpability is determined
    o Approach to shared risk/reward

2. Operational

The operational category includes items where the detail can be subsequently developed into OLAs and OLMs by process forums and working groups:

    o Responsibilities for development of OLAs and OLMs
    o Compliance with OLAs and OLMs
    o Sharing of the OLF
    o Inter-party dependencies
    o Inter-party responsibilities

3. Process

Processes are defined in the process model. The contracts should include obligations to comply with the process model (see section **3.1.4 Process models**):

    o Organization and governance
    o Issue resolution
    o Common toolsets
    o Common policies and procedures

Some service providers may resist accepting obligations that they perceive as being overly dependent upon others for delivery. It is very difficult to predict every possible situation when defining contracts. As well as careful design of the service groups, building collaboration requirements into contracts is one way to address these challenges.

3.1.3.4 Service integrator contract

The service integrator contract will differ from a service provider contract in a number of ways, including:

- The contract will typically have clauses that manage conflicts of interest and allow the service integrator to act as an agent on behalf of the customer organization
- The targets will focus on end-to-end performance and not just the performance of the organization acting as the service integrator
- There will be targets related to collaboration and improvement across the SIAM ecosystem

The detailed requirements in the service integrator contract should be in alignment with those in the service provider contracts, in particular, governance. There could be challenges if the service integrator has a specific governance requirement over the service providers and there is no corresponding commitment in the service provider contracts. The customer organization's expectations of the service integrator must also be reflected in the service provider contract.

The service integrator's contract should also be aligned with the particular SIAM model. Issues can arise if a standard contract from an external service integrator is accepted without a full review against what is required to support the efficient operation of the SIAM model.

A service integrator contract can use the components of the generic structure for SIAM described in the previous section. The schedules and SoWs would describe the specific services that the service integrator provides. If the service integrator is provided internally, an agreement structured in the same way will fulfil the function of the service integrator contract.

In a Lead Supplier structure, there can either be a single contract with one MSA, with separate schedules and SoWs for the service integrator and service provider activities; or one contract for the service integrator activities and a separate contract for the service provider activities, each with a common MSA. Having only one contract in a Lead Supplier structure can create challenges if termination is required for one of the activities that the service provider performs but not the other.

Accountability

In a bid briefing for an early service integration request for information (RFI), a potential external service integrator asked of the customer organization, *'How can the service integrator be accountable for the performance of the service providers?'*

The answer was, *'What good are you if you cannot?'*.

Although this answer may seem flippant, it is a call to action. How does the service integrator influence the performance of the service providers?

3.1.3.5 Service provider contracts

Service provider contracts typically have a similar structure to the generic structure described previously, including a common services schedule, supporting schedules, SoWs and a common collaboration agreement.

Any obligations and schedules that are the same for several service providers can be included in the contract structure using a common services schedule. SoWs can then be used to define the specific services and targets, with supporting schedules defining other unique requirements. This approach is easy to manage and change, transparent and fair. In some SIAM models the structure of contracts will vary according to whether the service provider is strategic, tactical, operational or commodity.

As well as specifying the specific requirements, a service provider contract in a SIAM ecosystem should:

- Contain the obligations information that aligns the service provider to the collaboration and end-to end process requirements. These are usually included in a common services schedule.
- Recognize the specific role of the service integrator as the agent of the customer, whether the service integrator is internally or externally sourced.

Common law

In common law systems (UK, Australia, Canada, USA, India and some other jurisdictions), under contract law, an agreement is between named parties. Other parties (organizations) have no role in it.

In a multi-service provider ecosystem, this poses a challenge. The customer organization owns the contracts with the service providers, but the service integrator is carrying out many activities on its behalf.

This disparity is resolved by the service integrator being appointed by the customer as its agent in certain defined areas. It is common to define limits to delegated authority and to withhold authority in some areas (for example, the final authorization of payment).

There are a variety of mechanisms that can be used to support service integration. The service integrator is responsible for maintaining standards, including any tooling or processes, with which the service providers should comply. The details of these are not in themselves part of the contract but reference can be made to them in the contract supporting their subsequent development and use (see section **3.1.2.3** on **Non-contractual arrangements**).

The contract must ensure that each service element is assigned to a specific service provider (see sections **2.5.5 Existing services and service groupings** and **3.1.1.1 Service groupings**). Where the same type of service is provided by two service providers, for example, where hosting is provided by two organizations to support service continuity, these must be treated as separate and distinct service elements within the contracts. This ensures that:

- Only one service provider (internal or external) is accountable for each service element
- All service elements have an assigned service provider and are included in the sourcing model

This is known as mutually exclusive; collectively exhaustive (MECE). Any form of assignment that does not satisfy these MECE conditions is likely to cause issues later on. Service providers tend to be very happy to be paid for work that others perform but are less happy to be held accountable for performance where others have a material influence on what happens.

Even if a service provider provides a number of different services in a service group, the individual services should be listed separately in the contract schedules and SoWs. This supports future cancellation or transfer of an individual service, provided that the contract allows for this. The recommended approach to the construction of a service provider contract is:

- Determine the initial, outline statements of work for the services to be provided.
- Determine the interfaces, interactions and dependencies between the service provider, other service providers, the service integrator and the customer, adding obligations around these to the service provider schedules as necessary.
- Ensure that the targets, processes, governance and other unifying mechanisms apply and are included in the schedules (using references where appropriate).
- Ensure that requirements to take part in collaboration are included, such as:
  - o Participation in the structural elements (boards, process forums and working groups)
  - o Review of another service provider's changes/release plans when required
- Include a mechanism for 'excusing causes', where a service level failure was because of another party.

Requirements will change over time. Targets and contracts cannot cover, nor anticipate, all future events. Each contract needs to include the ability to extend or change services and obligations to meet changing requirements and scenarios.

A contractual key concept is service accountability, where a service provider is commercially and legally accountable for compliance with all contractual obligations, including service level achievements. This is irrespective of whether it really understands its obligations or uses subcontractors to deliver its services. This needs to be defined, documented and understood at the outset of any contract. If not stated clearly, this could leave the customer organization without recourse when the service provider is not delivering the agreed outcomes.

Every contract negotiation should begin with a baseline set of expectations on both sides, to address:

- Scope
- Quality
- Services
- Cost
- Maintainability
- Governance
- Reporting
- Regulatory
- Personnel

The baseline should be sufficiently specific to begin negotiation and allow for decision making to commence, but fluid enough that terms can be reached on a variety of scenarios either at the start of the contract or over time. The baseline should be considered as the minimum requirements.

In a traditional sourcing contract, the emphasis in contractual definition typically focuses on:

- Financial model, including charges and service credits
- Targets for services delivered
- Scope of services provided
- Transition and transformation
- Liabilities, waivers, termination, breach and dispute resolution
- Further standard schedules including governance, contract change, exit and others

In a SIAM environment, these concerns should be augmented by:

- Coordination and collaboration mechanisms
- End-to-end service performance and enhancements targets

3.1.3.6 Service credits

It is important to retain a perspective on how service levels affect a service provider's priorities under the contract, and the consequences of non-performance (often referred to as 'service credits'). These questions should be considered.

- Damages for breach of contract are meant to put the innocent party (in this case the customer organization) in the same position as if the contract were properly performed. If a service credit does not fully compensate the customer, what is the advantage of a service credit over a claim for damages?
- Most contracts can be terminated for 'material breach', for example, a breach that has a serious or significant effect. If service credits provide remedies for certain types of breach, is it to be implied that those breaches are not material?

- Are service credits the sole remedy for certain breaches? Within a SIAM model it is far better to allow for the requirement for an offending provider to undertake a service improvement or perhaps some 'value add' and non-chargeable activity instead. This improves overall performance rather than simply restricting the profit of the service provider.

- Does the service credit regime lead to an implication that the customer is prepared to accept a certain amount of non-performance? Are service credits driving the desired behavior from service providers, or is it more cost effective for them to pay for poor performance rather than correct it?

- Would it be better to pay an incentive for good performance rather than a 'penalty' or adjustment for failure or partial failure?

- What is the consequence of a service provider's failure to comply with the service credit regime itself (for example, a failure of reporting, delay or invoicing irregularities)?

- What remedies can the customer apply if a service provider fails to meet an obligation that is not associated with service credits in the contract?

One way to align the service provider's performance to the interests of the customer is for the service level objectives and targets to directly support the customer's business objectives and goals. There are many ways in which service credits can be calculated, such as:

- Percentage rebates from the service charges for each percentage point that the service provision falls below the service level target.

- The use of service credit 'points' across a range of service level measures, which are then converted into service credits based on a formula over an agreed measurement period, usually on a monthly basis.

Service credits to promote service improvements

If a service provider is given an incentive to focus on rectifying the root causes of problems, service credit regimes might include mechanisms that impose multipliers on the service credits that are payable in the event that problems recur within particular timescales.

This is especially useful if the problems are trivial but annoying.

Service level and service credit regimes should incentivize performance under the contract. Unless the service level and service credit regimes are aligned with the objectives of the contract, there is a risk that the mechanisms can become a monthly administrative task, with relatively trivial amounts of money at stake. In these circumstances, service levels and service credits simply become a contract overhead with no real benefit to either party.

Properly constructed service level and service credit regimes can provide both early warning signs of problems in service delivery and financial incentives to rectify poor performance. To be effective, service level and service credit regimes need careful drafting, analysis of the circumstances in which the regimes will be applied and a good understanding of the legal/contractual environment in which they operate. These regimes deserve rather more attention than they are frequently given in the negotiation and drafting of contracts.

Legal status of service credits

When drafting contracts during the Plan & Build stage, country-specific guidance must be sought. It is often assumed that as service credits provide a pre-specified financial remedy in the event of poor performance, they are a form of liquidated damages. If so, in order for the service credits to be enforceable by the customer, they must not exceed a reasonable pre-estimate of the customer's likely losses in the event of poor performance. This can be quite difficult to establish, as there may be, in practice, relatively little financial impact on a customer if the service level target is missed by only a small margin.

An alternative approach that gives greater flexibility over the level of the service credits that can be levied is to make clear that the service level and service credit regime is not a form of liquidated damages, but is simply a mechanism that specifies that the customer will pay a defined service charge for a different level of performance by the service provider.

On this basis, the customer and the service provider have the freedom to enter into a contract for a varying service charge depending on the service levels that are achieved. This approach has the benefit that the level of the service credits is not constrained by a reasonable pre-estimate of the customer's likely losses if the service levels are not achieved. Although it is consistent with the preservation of the customer's common law remedies, the major drawback is that as long as the service provider's performance remains within the range defined by the service level and service credit regime, there may not be a breach of contract.

Service credit regime, part one

Failure to meet service levels in a particular month resulted in a service provider accruing service points, depending on the level of failure against a defined scale, up to a maximum amount. For example, the target for availability was 99% and 100 service points were allocated for every percentage point below the target. So, 100 points would be accrued if availability was only 98%.

At the end of each month, a percentage of the monthly service charge was retained, using a ratio of one percentage point reduction for each 100 service points. The service provider could earn that back if it achieved the availability service level for the next three months in succession.

The same approach was used for all other service levels.

Consider the situation where there is a continually poor level of performance by the service provider, but there is no mechanism in the contract to deal with this. In these circumstances, it will be difficult for a customer to assert that a 'material breach' has occurred that would entitle the customer to terminate the contract.

Service credit regime, part two

(continued from part one)

The maximum number of points that could be accrued in any month for failure to meet availability targets was 1,000.

> If this level was reached, it was called a 'critical failure'. This would result in a warning notice to the provider. Two consecutive months with 'critical failures' was defined as a material breach of contract, which could lead to termination.

Another approach is for the parties to agree a termination threshold based on the level of service credits that accrue over a particular period, to fit together with a right of termination if there are persistent defaults. In practice, however, it is not easy to decide upon an appropriate level for the termination threshold, and service providers will often seek to impose thresholds that allow latitude for failure.

An approach used for positive means is for the service provider to spend the value of the notional service credits on service improvement activities for the customer.

Capped service credits

Service credits are frequently 'capped' at an overall percentage of the monthly or annual service charges. In larger or long-term contracts, it is sometimes inferred that the service credits could remove the service provider's profit. However, given the focus of a SIAM ecosystem, working on this basis is likely to be counterproductive, especially if it causes the service provider to operate at a loss, as this is likely to lead to poor relationships and less collaboration.

A 'capped' service credit may appear to be advantageous to the service provider, as it has an almost guaranteed revenue stream despite its actual level of performance. However, if the contract also has a common law clause allowing the customer to recover its actual losses in the event of the level of performance falling below the service level and service credit regime, then the customer's position in these circumstances can be more advantageous than if there were an uncapped service credit mechanism.

Towards the end of a contract, it may be cheaper for a service provider to repeatedly receive service credits rather than invest to prevent them, for example, if its technology infrastructure is at end of life. The customer organization should seek to avoid such 'gaming' behavior by building end-of-life clauses into the contracts.

3.1.3.7 Incentives

The aim of designing service payment models is to reward service providers fairly for effort expended and to align interests to the greatest possible joint benefit. Most simple models suffer drawbacks, for example, the payment of a service desk on the number of incident calls logged can motivate a service provider to maximize incident record numbers.

It is often desirable to design a model that first provides fair recompense, and then overlay a second level that aligns incentives. This is an inexact science and there is no one right way to do things. Take care to avoid overcomplicated models that may become too complex to implement and administer. Perfection is of little use when it is impossible to reconcile a monthly invoice to actual activity and performance.

Interpretation of incentives

A service desk was paid an additional amount depending on the number of service 'compliments' it received each month. The customer organization hoped this would incentivize the service desk provider to focus on truly excellent service.

The number of compliments logged was very high, but anecdotal evidence suggested that customer satisfaction was not, in fact, very good. On investigation, the customer organization found the 'compliments' logged were not what they would recognize as a compliment – for example, a customer emailing 'thanks' in response to a service desk email.

The terms built into the contracts need to be reviewed regularly to ensure they are delivering the intended results and to identify improvement opportunities.

Ill-judged incentive schemes can quickly promote perverse behavior. There is a managerial cliché that suggests that if you wish something to be taken seriously, measure and reward upon it, indeed, this works. Just be careful what is measured. Many organizations will make mistakes and suffer consequences to service quality. Construct something with huge complexity and there is more room for argument. The most probable outcome in this case is confusion followed by inertia.

The general guidance is to be moderate in the application of incentives. Try something and, if it works, consider a modest extension. Collaborate when setting targets, allow service providers to know what is important and what the customer organization is trying to achieve.

In the context of the design of a SIAM model, a subject that frequently occurs is that of incentivizing cooperation and collaboration. Given the criticality of these behaviors, it is natural and wise to seek a variety of leading and lagging indicators and to attach value to these. The chief lagging indicator (measure of outcome) is commonly 'end-to-end service availability' or 'major incident recovery time' (see section **3.1.8.1 Coordination mechanisms**).

There can be opposition from service providers if excessive weight is given to incentives where an individual service provider's reward is dependent on the behavior of others. The appropriate response is not to discard the mechanism but to apply it modestly so that it remains an incentive to exhibit the desired behavior but does not form a core component of the service provider's base service payment.

Incentives should look beyond payment alone. It helps greatly to know and consider how the other party's reputation is managed within its own organization. If the customer organization or service integrator is generally cooperative and helpful, the impact is all the greater when behavior or performance means that a reprimand is sharply administered.

Reputation

In one organization, a global service provider, if it received a warning notice for any of its contracts, the chief executive had to inform global headquarters, provide then with an action plan, and report weekly on progress to both the customer and headquarters.

Ensure that there is a sensible mechanism in the contract to deal with non-achievement of a requirement. Service levels and credits are easy, but what happens if a service provider repeatedly fails to provide a service report? Unless there is a specific mechanism, all the service integrator can do is treat it as a material breach of contract and advise the customer organization to terminate the service provider, which is not necessarily the ideal way to manage the situation.

3.1.3.8 Intellectual property

Intellectual property (IP) is a commonly misunderstood area with the capacity for unpleasant surprises and dispute. Customers frequently assume that they have the right to access and use everything even vaguely connected with their service. However, in many countries, the legal rights to artefacts (for example, a design or a process) rest by default with the creator (or author). In a SIAM model, the creator is often another organization.

Outsourcing contracts commonly make provision for software licencing assignment, but do not always address areas including process, policy, design, operating methods, scripts, application maintenance tooling, knowledge articles, recovery instructions, operating schedules and the like, all of which are required to transfer a service to a different service provider. If the exit schedule and IP provisions within the outgoing service provider's agreement are inadequate, this can create some challenges during the transition (such as unwillingness to transfer or additional charges).

What a customer organization needs is the right to use IP and associated artefacts, where particular to its service, including the right to make this available to an incoming service provider. This can be included in the contracts with external organizations.

Shared operations centers

It is common for service providers to operate elements of their services through operations centers that serve many customers. They may not want to share their IP with other providers of one particular customer. Sharing information with a competitor's shared operations center may provide the advantage necessary in a subsequent contest for business.

After all, an incoming service provider's operations center will not be visited by an outgoing service provider, but an outgoing service provider's operations center will be visited by the incoming service provider.

If a customer leaks a service provider's IP to another provider, it is a violation of trust and may expose the customer organization to a claim for damages.

3.1.3.9 Dispute management

There must be defined procedures for managing disputes between the different parties involved within the SIAM ecosystem. Disputes may occur between the customer organization and one of the other parties, between the service integrator and a service provider, or between different service providers. Whatever the case, there needs to be a defined and agreed procedure that aims to resolve the dispute as amicably as possible, but allows escalation where necessary to achieve resolution.

Settling a dispute by legal means will ideally be a last resort, particularly in a SIAM model where significant focus should be given to developing a culture of partnership and collaboration. Without defined procedures for handling disputes, there will be a tendency for minor disagreements to either linger without resolution, gradually creating increased bad feeling and preventing true partnership, or for the disagreements to escalate unnecessarily resulting in legal action that could have been avoided.

Contracts need to ensure that:

- Dispute management procedures are defined, documented and agreed
- Roles and responsibilities regarding dispute management are clearly defined and allocated
- Compliance with defined dispute management procedures is made part of service provider contracts
- Dispute management procedures actively seek to find amicable solutions that are fair to all parties
- There are clear mechanisms for triggering dispute management
- Disputes are recorded, owned and tracked through to resolution
- Legal measures are taken only as a last resort

3.1.3.10 End of contract

When the need arises, notice of termination of a service provider will be formally issued some time in advance, in accordance with contractually defined procedures. It is important to get this right and good legal advice is essential, particularly where the situation is contentious. The final closure of the service should be documented to ensure that there is minimal scope for later dispute. The closure should address:

- Stranded cost (see below)
- Settlement of debt
- External party obligation and agreement transfer
- Staff
- Assets
- Agreeing the point at which the obligations cease
- Inflight projects
- Access, permissions
- The disposal of confidential information

Stranded costs

Stranded costs represent assets that may become redundant after a substantial change, such as a move to a SIAM model. This is normally provided for in an exit schedule.

For example, a service provider supported a packaged application, the level three maintenance of which was sourced by the package provider on an annual maintenance agreement. The service contract with the service provider was to end shortly (in September), but the anniversary for the package maintenance contract was in April. As such, at the time of termination the customer had consumed five months' worth of service, but another seven months had been paid for by the service provider.

The service provider faced the prospect of not recovering this expense from the customer, the cost being 'stranded'. Under the exit schedule in the agreement with the service provider, stranded costs were recoverable as loss.

Avoid a 'zombie contract' in which most obligations to provide services cease but some remain. Some, such as enduring obligations to maintain confidence, should be maintained. Much beyond this should be avoided.

3.1.3.11 Exit services schedule

It is common practice for service provider contracts, including any external service integrator, to contain an exit services schedule that prepares for the eventuality of service exit.

As this schedule is created before the first service has been delivered, it may not be fully complete where details are not known. Without robust exit plans, it can be extremely difficult for a subsequent incoming service provider to pick up the service with the information available, so the schedule should be as complete as possible and maintained as changes occur.

3.1.4 Process models

In a SIAM model, the execution of most processes will involve multiple service providers. Each service provider might carry out individual steps in a different way, but as part of an overall integrated process model (see section **2.7.2.3 Managing process outcomes, not activities**).

A process model shows process and service relationship interactions with the customer organization, the service integrator and service provider teams. A continual improvement mechanism should be built into the model. A process model could use an available industry framework, such as the ITIL Continual Service Improvement (CSI), Business Process Improvement (BPI) or other relevant approaches.

Operational process activities

Process models help to make it clear who does what within the SIAM model. There is a common misconception that the service integrator will take care of all the process activities, this is not the case.

Some customer organizations receive service integrator bids that seem to be low cost. This can be because of the customer assuming the service integrator will carry out all operational process activities, and the service integrator understands that it will only carry out process integration.

Process models are important SIAM artefacts. Individual processes and work instructions are likely to remain within the domain of the individual service providers. The process model for each process should describe:

- Purpose and outcomes
- High-level activities
- Inputs, outputs, interactions and dependencies with other processes
- Inputs, outputs and interactions between the different parties (for example, between the service providers and the service integrator)
- Controls
- Measures
- Supporting policies and templates

Techniques such as swimlane models, RACI matrices and process mapping are commonly used and are helpful for establishing and communicating process models. The process models will continue to evolve and improve as further activities are undertaken during the Plan & Build stage, and then once live in the Run & Improve stage.

Techniques such as process modeling can be used to explain the required activities more simply. Process modeling is the analytical illustration or representation of an organization's activity or set of activities intended to support the achievement of specific goals or outcomes.

In the Discovery & Strategy stage, any existing processes are mapped to create a baseline for process improvements and integration (see section **2.5.4 Current organizational capability**). Within the Plan & Build stage, process modeling allows the design of new or 'to-be' processes to address any gaps.

3.1.4.1 Integrated process models

Within a SIAM model, process execution is likely to involve multiple stakeholders. Despite this, it is not necessary for all service providers to use the same process documents or tools. To preserve the economic specialism and expertise of individual service providers, the service integrator needs to provide the appropriate inputs and define the expected outputs and outcomes. Each service provider might carry out individual steps in a different way, but as part of an overall integrated process model with defined interactions, rules and controls.

ISO 9001:2015 (for quality management systems) expects organizations to adopt a process approach that supports the management and control of processes, the interactions between processes, and the inputs and outputs that tie these processes together. This is the kind of approach that is required within a SIAM model, and the service integrator is responsible for managing the processes and process interactions as a coherent whole based on the agreed governance and quality management in place.

Outputs are the result of a process. Outputs include not only services, software, hardware and processed materials, but also decisions, directions, instructions, plans, policies, proposals, solutions, expectations, regulations, requirements, recommendations, complaints, comments, measurements and reports.

The output of an upstream process often becomes the input for a downstream process. The service integrator should define the controls that enable process outcomes. Where possible, interactions should be automated to provide consistency and reduce manual activity.

Within a SIAM model, multiple service providers could undertake multiple process activities, leading to multiple possible input-output relationships that tie these processes together. To manage this complexity, it is suggested that processes are visualized one at a time using a single flowchart or a single page, mapping the interactions between the stakeholders for that process. This will allow the most important input-output relationships to be specified without getting distracted by complexity and detail. **Figure 16** shows, in general terms, how this can be achieved.

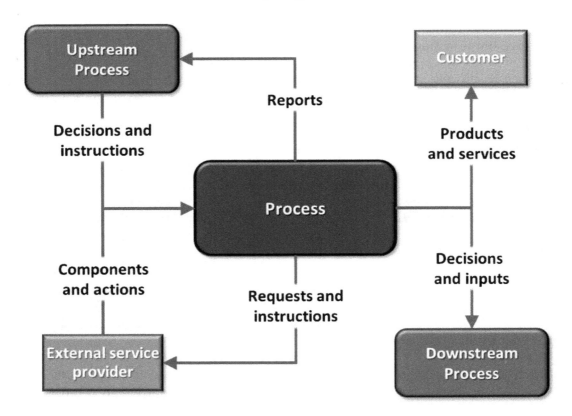

Figure 16: Process flowchart

The center box in **Figure 16** identifies the process being described, receiving instructions from upstream processes and providing inputs to downstream processes. Arrows represent inputs and outputs, as described by the associated text.

This type of process map allows a high-level view of the stakeholders involved within the process interaction. Further, more detailed process flows may be required. These can be adopted and adapted from the service management framework of choice. For example, service providers using frameworks such as ITIL or conforming to standards such as ISO/IEC 20000, will be familiar with terms such as policy, process, procedure and work instruction. Within a SIAM model, policy and process models are defined and mandated as part of the overall governance framework, and are common to all, where possible. Procedures and work instructions are proprietary to service providers. The service integrator should focus on outcomes, rather than forcing service providers to work in a particular way.

Lack of collaboration and trust

In a recent SIAM transition, the external service integrator requested that all service providers submit copies of their process documentation, role descriptions and performance metrics. There was no context given for how this information was to be used.

Initially, all the service providers refused. The service integrator was considered to be a competitor and the service providers' process documentation, role descriptions and individual contract details were considered 'commercial in confidence'.

It took another ten weeks to form sufficient trust for the service providers to fully contribute to the process model design and release the required information, in a controlled manner. If the project had set up appropriate governance to handle some of these sensitivities, this initial challenge – and delay – could have been avoided.

The service integrator will set the minimum standard for documented information. However, it is good practice to allow individual service providers to:

- Provide and maintain process documentation to the extent necessary to support the operation of processes
- Retain documented information to the extent necessary to be confident that processes are being carried out as planned

Service providers should share documentation that shows the interactions for each process between the parties with the service integrator (and vice versa), and with other service providers.

This approach provides for some flexibility. It is good practice for the service integrator to use flowcharts to give other participants within a process a view of the big picture. It allows the individual service providers to develop more detailed procedures to show how the process activities should be carried out, and at which point they will be required to contribute. Smaller, less mature service providers may benefit from a more prescriptive mandate around process flow from the service integrator. Conversely, it is unlikely that a large cloud service provider would accept a prescriptive process flow or mandated toolset as part of its obligations.

Macro and micro processes

One solution is for the service integrator to provide a macro process, for example, change management. Each macro process contains a number of micro processes.

The macro process shows an overview, has an owner and contains the policies, goals and metrics. The micro processes are prescribed procedures such as normal change, emergency change or standard change. The micro processes are modelled in detail and each activity is linked to roles.

Some service providers may adopt the micro processes, whereas others will use their own. As long as the interactions, inputs and outputs between service providers and the service integrator meet the prescribed macro process, the service integrator retains consistency and control across the end-to-end process.

3.1.4.2 Managing process outcomes and integrating processes

No process exists or can function in isolation. It is essential to clearly document and ensure understanding of the required inputs, outputs, interactions and dependencies of each of the cross-provider processes. The service integrator should also communicate appropriate outputs for certain processes, for example, demand forecasts for the capacity management process.

It is not necessary for the service integrator or service providers to understand the minutia of each other's process. Each party, however, needs to understand the expectations of each interaction in terms of response time, data requirements, communication paths, inputs expected and outputs to be delivered. The service integrator needs to be able to provide appropriate inputs, receive agreed outputs and provide a feed into continual service improvement activities.

Standardized roles and process interactions

One of the goals of a SIAM model is the efficient management of complex supply chains. Standardization and automation can support efficiency.

A good practice is for the service integrator to provide the service providers with standardized role descriptions, covering the requirements necessary for execution of the integrated processes. Each service provider can then map those roles to how they operate internally. This simplifies understanding of how to engage with the roles in the service integrator and other service providers, and supports consistent terminology.

In one case study, a set of standardized roles was defined that included a process manager for each process. Some service providers elected to allocate the same contact point for all processes, where others assigned different contact points. However, by using the standard role descriptions (and mapping), it was easy to understand who to contact for any process – across all service providers.

Pay particular attention to handovers between parties. For example, does one service provider assume that the other will provide data that the other does not think is an output? Or, does the provider of the data require confirmation that it has been received and understood?

Street versus house level change

One area in which many customer organizations struggle to relinquish control is complete ownership of change management. Moving to an outcome-based service may include a change to a cloud-based leveraged solution, where the customer organization is one of many and has no control of change management.

This can be addressed in a SIAM environment by categorizing 'street' level change and 'house' level change. Anything that occurs at the 'house' level is subject to change control by the individual service provider. Anything occurring at the 'street' level requires attention from the service integrator (in conjunction with one or more service providers and/or the customer organization, as illustrated in **figure 17** and **figure 18**).

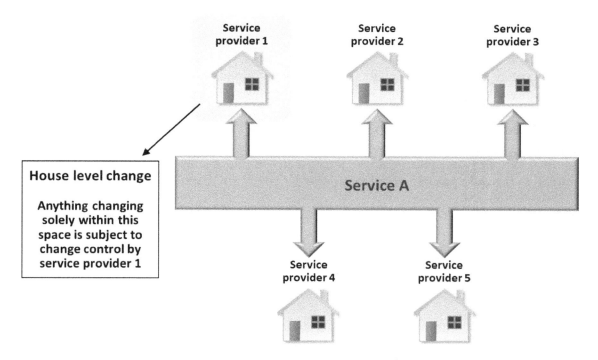

Figure 17: House level change

The analogy can be expanded with a '**neighbor change**' between two parties that are tightly related but with no further impact.

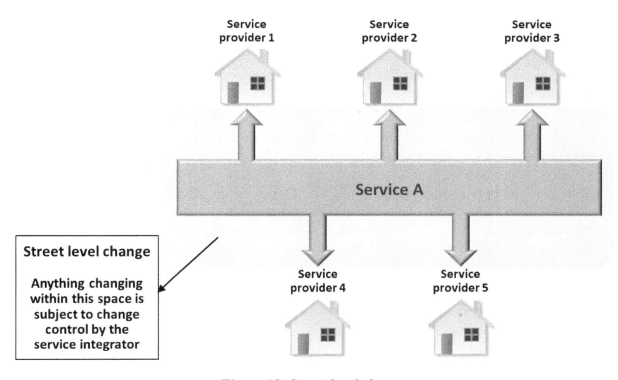

Figure 18: Street level change

Street versus house

This example can be used to explain the scope and impact of each service provider. Imagine returning from work and finding that the lights in your house were not working. When you pressed the switch, nothing happened. One of the first things you might do would be to go outside and see if other houses in your street (or apartments in your block) were experiencing the same issue.

If the issue is only in your house, you might contact your electrician first. If the issue did affect the whole street, you would probably contact your electricity provider first.

3.1.4.3 Operations manual

The operations manual (sometimes referred to as a run book) is typically created to provide a summary of the contract, service, deliverables and obligations in layman's terms, and to support an understanding of process activities and objectives.

The operations manual communicates how individual service providers will engage with others and with the service integrator. It is a useful tool for developing and documenting an understanding of the goals and objectives of the customer organization, and how those goals and objectives cascade down to relevant service providers. It outlines the policies, procedures and service levels required, and how processes interact between stakeholders. In terms of output content, format and timescales (for example, operational level agreements (OLAs)), the operations manual represents guidance only and is not intended to create any legally binding duties or obligations.

3.1.5 Governance model

The Discovery & Strategy stage will provide guidance regarding the defined systems of control (see section **2.3 Establish a SIAM governance framework**). In the Plan & Build stage, SIAM governance focuses on three key aspects:

1. Ensuring alignment between the SIAM strategy and the current and future needs of the customer organization
2. Ensuring that the SIAM strategy and SIAM model are planned and implemented successfully
3. Ensuring that the implemented SIAM model can be managed and operated in a controlled and collaborative manner, being both efficient and effective

The governance model should be designed based on the governance framework, roles and responsibilities. It includes scope, accountabilities, responsibilities, meeting formats and frequencies, inputs, outputs, hierarchy, terms of reference and related policies in force within the SIAM model. For each governance board, this model should include:

- Scope
- Accountabilities
- Responsibilities
- Attendees, including chair and secretariat
- Meeting formats
- Meeting frequencies
- Inputs (and who is responsible for them)

- Outputs (including reports)
- Hierarchy
- Terms of reference
- Related policies

The challenge when designing a SIAM model, is taking the complex governance structures and ensuring that these are enabled through Plan & Build and into the subsequent roadmap stages. The rules, relationships, policies, systems and processes that define authority levels within each organization must be defined, clearly stated, exercised and maintained.

The governance framework provides the guardrails that control Plan & Build activities and outcomes, helping to drive enhanced organizational performance while at the same time aiding conformity with business requirements (for example, the customer organization's constitution, policies, controls and procedures, as well as with applicable external regulations and laws).

Managing out-of-scope service providers

In some cases, not all service providers fall within the controls of the SIAM model, for example, other external organizations and internal functions that contribute to delivering services. Some organizations (for example, cloud service providers) only provide commodity services and do not engage with the SIAM environment. Some service providers are part of a group, where the relationship is with the parent company, so there is no direct control.

Where this occurs, it is still necessary for the service integrator to understand and document the interactions with these service providers and, where necessary, act on their behalf (for example, looking at the cloud service provider's website for any planned upgrades). This will enable the customer organization to have the complete view of service provision necessary for appropriate governance.

To realize the full value of implementing a structured governance model as part of the SIAM model, these considerations are important:

- Measure performance at a level appropriate to model maturity
- Align service provision to business objectives
- Take an Agile approach – start small and develop through iterations
- Create a tooling strategy to provide a 'single source of truth' for proper governance and reporting

Compliance management and audits

Within the Plan & Build stage, the roles and responsibilities of the service integrator for compliance management and audits must be clearly outlined.

Typically, these include:

- Record keeping (as an ongoing activity and in preparation for external audits)
- Verification of service provider records
- Ensuring adherence to all compliance norms
- Ensuring compliance parameters are followed during regular change management

- Ownership and accountability of all service providers meeting compliance requirements
- Ownership and accountability of maintaining compliance within the service integration layer
- Tracking and providing regular compliance reports with clear actions to address any potential/identified issues
- Ownership of performing and completing root cause analysis on any identified compliance issues
- Clear understanding of all regulatory/compliance requirements as well as all audit parameters
- Implementation and oversight of operational systems to ensure ongoing compliance adherence
- Ensuring all systems and processes conform to critical requirements, such as segregation of duties
- Ensuring that SIAM processes and systems follow the habit of separating through an 'ethical wall', where there are clear guidelines on the sharing of information pertaining to each service provider with other service providers
- Monitoring alignment with corporate governance requirements, for example, sustainability, visas, etc.

Audit schedules should be defined that plan the approach to assurance. These should focus on assuring delivery in line with compliance and quality requirements. They should also focus on areas with potentially high impact for non-compliance or poor performance. Audits can be undertaken either horizontally, for example, across a SIAM layer; vertically, by functions or processes; or organizationally, such as the service integrator or a service provider.

Audit schedules typically include information:

- Identifying the scope of audits over a timeline, typically a 12-month period, based on business requirements
- Detailing the processes or functions to be audited and assigned auditors
- Identifying the timeline planned to complete the audits
- Scheduling the management meetings following an audit to review findings

3.1.6 *Detailed roles and responsibilities*

In the Discovery & Strategy stage, the key principles and policies for roles and responsibilities are created (see section **2.4 Define principles and policies for roles and responsibilities**). These should include the detailed design and allocation of roles and responsibilities for:

- Process models
- Practices
- Governance boards
- Process forums
- Working groups
- Organizational structures and locations for any retained capabilities

This work may highlight a need to review earlier design decisions.

Roles and responsibilities can be further developed or evolve in the Run & Improve stage, but the details must be confirmed in the Plan & Build stage before any service integrator or service providers can be appointed. There are three aspects about which every member of the SIAM ecosystem must be clear:

- The function or role to which it reports
- Its responsibilities and corresponding expectations
- The level of authority it requires to make decisions, and its level of empowerment to act

---

A role is not a responsibility, and vice versa

- A role is: *"the position or purpose that someone or something has in a situation, organization, society or relationship."*[20]
- A responsibility is: *"something that it is your job or duty to deal with."*[21]

If the organization acting as the service integrator also takes on a service provider role (as in the Lead Supplier structure), the management structures and personnel acting within the two elements need to be distinct (see section **2.3.9 Segregation of duties**). Otherwise, the impartiality of the service integrator can be questioned, with the perception that its 'own' service provider is being treated more or less favorably.

It is highly advantageous, however not always possible, for internal teams (service providers) to be treated in the same way as the external service providers within the SIAM model. For example, by using an operational level agreement (OLA) to map obligations of and interactions with these teams. This makes it easier for the service integrator to manage across all service providers, and avoids any concerns relating to internal service providers getting preferential treatment by the service integrator or the retained capabilities. If this is not possible, the impact of treating them differently needs to be understood and considered in the SIAM model.

Some (external) service providers may maintain a direct relationship with a part of the customer organization outside of the service integrator (for example, a payroll application service provider providing services to the finance department). As these service providers are still a part of the SIAM model, they should also be managed by the service integrator like any other service provider. If these direct relations need to persist, the service integrator may need to carefully consider how to create an inclusive forum where they can still manage the end-to-end delivery.

Capability framework and skills maps

The nature of a SIAM model requires people to demonstrate or develop very specific skills. With the complexity of the various cross-functional and cross-organizational teams working within the ecosystem, soft skills such as negotiation and influencing are important.

For example, it is essential for staff working within a 24/7, global SIAM model to develop cross-cultural competencies as they hand over instances such as incidents, changes and service requests across countries and time zones. Similarly, collaboration skills (such as communication, negotiation and mediation) may be of specific importance for certain roles in a SIAM model.

---

[20] *https://dictionary.cambridge.org/dictionary/english/role*.

[21] *https://dictionary.cambridge.org/dictionary/english/responsibility*.

Role descriptions within the SIAM model

A bad practice was observed in a global engineering enterprise. After the target SIAM model had been agreed, it was not communicated effectively to an external consultancy that had been hired to define the organizational structure and job descriptions.

As a result, the consultancy used 'best guess' assumptions to draft the job descriptions of the retained capabilities. These were not aligned to the target SIAM model, creating confusion, delay and rework of the roles and responsibilities. Since no skills framework had been used, the roles had not been levelled and normalized between departments, creating overlapping responsibilities, irrelevant skills and imprecise capability descriptions.

This caused the HR department problems in trying to apply the correct grading and proceed with timely hiring. The team lost valuable time in the transition, and most job descriptions had to be rewritten 'on the fly'.

A capability framework provides a set of principles that clearly outline the expected behavior and capabilities required of staff to achieve performance. Based on this, each service provider, including the service integrator, will need to maintain a framework and systems for assessing the effectiveness of its people.

A skills map sets out all the main employability skills in a way that shows their relationships to each other. The innermost oval has primary or fundamental skills, the middle circles contain intermediate skills and the outer circles very specific skills. For example, communicating can be divided into speaking and writing, and spoken communication can be further broken down into presenting, listening and telephone skills (see **figure 19**).

Figure 19: Communication skills map

Having identified the levels of capability and capacity service providers need to deliver their services, the service providers will need to consider the skills they have against what they need. This will identify gaps evident by not having the right level of capability available at the right time.

3.1.7 Performance management and reporting framework

The performance management and reporting framework should provide valuable information to all stakeholders. It should be regularly reviewed and include metrics to provide information on a number of key areas, including:

- The performance of each service provider in the ecosystem against its targets
- The delivery of each service provider against its (contractual) obligations
- The end-to-end delivery of services across the ecosystem

The scope includes all internal and external service providers.

Content

The performance management and reporting framework addresses measuring and reporting on a range of items, including:

- Key performance indicators (KPIs) – these should be specific, measurable, agreed, relevant and time bound (SMART)

- Performance of processes and process models
- Achievement of service level targets
- System and service performance
- Adherence to contractual and non-contractual responsibilities
- Collaboration
- Customer satisfaction

Measurements should be taken for each service provider and its services, and also across the end-to-end SIAM ecosystem.

Designing an appropriate performance management and reporting framework for a SIAM ecosystem can be challenging. It is usually straightforward to measure the performance of an individual service provider. The challenge is in measuring end-to-end performance as experienced by the users, particularly when there may be limited consistency in how each of the providers measure and report (see section **3.5.2 Measurement practices**). To avoid providing an inaccurate view of performance, the focus should be on measuring the quality of the services as experienced by users and customers.

The performance management and reporting framework should also include the standards for:

- Data classification
- Reporting formats and frequency

Measure many, report on few

Many organizations are moving from standardized weekly and monthly reports to exception-based reporting. This reduces the effort associated with reporting, as reports are only produced when the unexpected happens. If an organization moves to exception-based reporting, it still needs to be able to identify trends where a service or process is trending downwards towards an exception – measure many things, report on few.

It is important to be clear and concise and not to overwhelm the audience with detail. A common mistake is to make reports highly detailed (sometimes this level of reporting is requested by the customer organization). The principle should be to keep reports simple and practical to show overall performance, but then allow the recipient to request more detailed information if they need to. Identifying appropriate tools to allow the right level, structure and medium of reporting is an important consideration (see section **3.1.9 Tooling strategy**).

Where possible, it is important to keep a consistent format for trend analysis purposes.

There is a skill in reporting, which is an appreciation of what the audience is interested in. This extends beyond a reactive response to reporting requests (or contractual obligations), to anticipation of what the governance body should have asked for. A good report contains all that is necessary and no more, but can also recognize good performance and thank the staff responsible. Recognition is sometimes under-used, but can be a powerful motivator when included in reporting (to a higher level in the organization).

Stakeholders

The main stakeholders of the performance management and reporting framework include:

- Customer organization
- Senior management team
- Purchasing and/or finance
- Business relationship managers
- Management of service provider organizations

Although the end users of the services are also stakeholders, the performance management and reporting framework is designed primarily as a management tool for the customer organization. Many organizations use a summary of the information and reports derived from this framework to inform the end user community of service performance.

External provider obligations

It is important to distinguish between performance (how well something is being done) and delivery (of contracted commitments and the fulfilment of all obligations). Each provider will have a contract that will specify obligations (delivery) and KPIs/SLAs (performance).

Example obligations include:

- Responsibilities of the service delivery manager based on site
- Billing to be submitted by the fourth working day of the month
- Use and change of subcontractors to be notified to the customer two months in advance and subject to approval
- Attendance at and input to a collaboration/innovation forum

Typically, obligations last for the full term of the service provider contract and tend to be fixed unless updated by agreement. In some instances, not all obligations have to be fulfilled before the delivery of services can commence. Sometimes, not all requirements and possibilities are fully understood, and a period of learning is necessary. During the design of the performance management and reporting framework, it is important to define when obligations are expected to be fulfilled.

During the Implement stage, service providers should concentrate on meeting obligations that are key to delivery, such as setting up facilities, gaining knowledge and establishing processes.

Once in the Run & Improve stage, the priority should be to meet obligations that affect the ability to govern the services, such as producing performance reports, attending governance boards and other key structural elements.

Some obligations may not be able to be fulfilled without the customer organization or the service integrator taking action. Where this is the case, the service provider cannot be held liable for any delay in meeting its obligations if another party is causing the delay. Other obligations may be delayed because of their nature. For example, a requirement to provide a three-year summary of service statistics would not be possible until year four of the agreement.

Some obligations are event driven, not time dependent. For example, a technology refresh forum might only be required when a strategic change is requested, a technology refresh is due, or a technology becomes incompatible or obsolete. Attendance at a dispute resolution meeting is only required when a dispute occurs.

It is important to track obligations met and failed. The service integrator is responsible for tracking the status of the service providers in meeting obligations. The retained capabilities within the customer organization are responsible for tracking the service integrator's status similarly.

For each service provider, the fulfilment of all obligations should be monitored and progress recorded. This should track the obligations and follow the reporting principles mentioned above. A percentage of obligations met is a good summary indicator to provide to the customer organization.

Service provider performance

Performance is commonly measured through key performance indicators (KPIs) and service level agreements (SLAs).

Example performance measures include:

- Telephone calls to the service desk are answered within 30 seconds
- A major incident management working group is set up within 20 minutes of a major incident being declared
- Configuration management database (CMDB) records are 90% accurate
- Emergency changes form less than 5% of the total changes
- Less than 5% of releases are rolled back

KPIs, SLAs and OLAs for individual service providers are usually managed through the service level management process (see also sections **2.3.13 Supplier and contract management** and **3.1.3 The importance of contracts in SIAM**). However, end-to-end SLAs and KPIs require an additional level of management and reporting by the service integrator.

3.1.7.1 The role of targets

Well-defined targets set expectations and promises for the delivery of outcomes and resources. They support strategic objectives and resource allocation. Effective targets require a clear understanding of the link between inputs and their contribution, and the outputs and the business outcomes they support. They must be reviewed regularly to ensure they remain appropriate and aligned to business needs.

Targets will be scrutinized by potential service providers, as they may have to adjust their internal delivery models and approach to ensure they can meet them. This can sometimes adversely affect costs and other requirements that do not have targets associated with them.

Service providers will also need to assess any subsequent changes to targets and outcomes to see if they are achievable, resources are adequate and costs can be recovered. Unless the customer is prepared to fund any necessary changes, this can adversely affect the relationship between the service provider, the service integrator and the customer. Hence, any changes to targets require an informed impact-based discussion.

Impact-based discussion

'Yes, this could be done with half the peak cash flow in year three, but because of the need to review and probably change the interactions with service provider A and the service integrator, it would

---

> *affect the risk to improvement project X, which could take twice as long, at three times the overall cost.'*

With the number of organizations involved in a SIAM ecosystem, it is important to have a commonly agreed method of calculating measures, performance, outcomes and targets, including the data sources used. This can avoid arguments between parties that may have used a different dataset or a different calculation to show their desired outcome.

3.1.7.2 Develop a performance measurement plan

To measure actual performance against the set targets or benchmarks effectively, a plan must be conceived for collecting and analyzing the necessary performance data or information. This plan must describe the methods and techniques of collection and analysis, and the frequency of collection.

It also needs to clarify and confirm the roles and responsibilities for each of these tasks. The service providers will be responsible for providing measurements of specific service elements that fall within their remit. The service integrator is responsible for aggregating these to provide operational metrics to allow an understanding of service performance as well as an abstraction of the information for customer-facing end-to-end performance reports.

The plan should assess the availability of the data sources, the feasibility of the collection methods and identify any potential problems. It is a good idea to focus on or start from existing performance data and information where this exists.

A key principle is to start with a set of minimum viable reports. These provide the least amount of information necessary to understand the performance of the services. A minimum set will be more easily understood, and faster to implement than a more comprehensive set. The minimum viable reports can be used to build understanding and collect feedback before being expanded if necessary. Without this approach, many organizations generate more information than is actually required.

Information is most effective when it is visual and easy to understand. Using service dashboards and scorecards will increase the impact of reporting. A picture can be easier to understand than a long report, but it must be clear what each visual is and what it indicates. **Figure 24** shows an example of a Kanban board that can support visual management.

The performance measurement plan must also determine who is responsible for gathering, analyzing and reporting on performance data or information. This might be based on the logical fit of these responsibilities with the SIAM layers, looking at existing workload, capability, ownership and timeframe. **Table 4** provides an overview for capturing a performance measurement plan.

Table 4: Overview for a Performance Measurement Plan

Outcomes	Reflect those outcomes outlined in the results chain
Performance indicators	Select both quantitative and qualitative indicators, with these criteria: 1. Valid 2. Relevant 3. Reliable

	4. Simple 5. Cost-optimal
Data sources	Utilize existing data sources where possible
Collection methods	Identify collection methods such as surveys, interviews, etc.
Frequency	Describe how often you will gather the performance information
Responsibility	Determine the person(s) responsible for gathering, analyzing and reporting the performance information and data

3.1.8 Collaboration model

Collaborative relationships are necessary within a SIAM model because of the number of organizations involved. A collaborative practice involves teams and functions across the SIAM layers working together to achieve shared goals. Collaboration is achieved when these various parties develop mechanisms – structures, processes and skills – for bridging organizational and interpersonal differences, and together arrive at valuable outcomes.

This section refers to the collaboration model not as an artefact or template that can be copied, but as the combination of techniques by which the objectives of collaboration between the parties are achieved. These objectives are likely to include:

- Coherent end-to-end service delivery, drawing effectively upon the service providers involved.
  - o This requires the contracts in place to have collaboration at their core, which may require some contracts to be rewritten or updated. It may mean the introduction of a common set of metrics across all service provider contracts. Ideally, a standard collaboration addendum can be developed which, once accepted by all parties, can be added to the existing contracts. Experienced advisors or external support may be helpful here.
- Positive and effective interchanges between the service providers, the service integrator and the customer organization.
  - o Consider developing a set of 'softer' metrics, focused on rewarding positive behavior. For example, have incident, problem, change, development and project teams rank the service providers based upon their perception of how collaborative and helpful they have been.
  - o Ask service providers to praise one another for positive collaboration experiences, and target them for the frequency in which they do so (for example, a minimum number of 'praises' raised per month).
  - o Create a clear scope for the service integration layer. It is essential to define the scope of the service integrator role so that it is clear where accountability lies, including the key considerations, the degree of authority it possesses and the level of interaction with service providers and the customer organization. One way to determine these is through scenario-based testing, which will reveal the key roles required to manage common situations such as the resolution of a major incident.

- Rapid identification and resolution of inhibitors to performance and service delivery.
  - Survey service providers to ensure that they are satisfied with the working practices. Give them an opportunity to build and improve collaboration in the processes in which they are active. This can be achieved through the appointment of process owners and creation of process improvement forums where all service providers can participate.
  - Create clarity between the SIAM layers. Ensure service provider teams understand the part they play in the delivery of business services. Instigate 'back to the floor' sessions where they spend time in business departments.

There is no one perfect arrangement by which collaboration can be achieved. See section **3.1.8.1 Coordination mechanisms** for examples of techniques that may be useful.

The SIAM ecosystem and the relationships between the customer organization, service integrator and service providers create a unique environment. From sourcing and contractual negotiations, through to governance and operational management, there are specific SIAM considerations that must be factored in at the Plan & Build stage.

The cultural aspects of a transition to a SIAM model are one such consideration. An effective SIAM ecosystem is underpinned by relationships and appropriate behavior. Collaboration and coordination are key elements that drive a positive culture. The aim is for all stakeholders to work together in pursuit of a common goal.

The foundation of collaboration can be built and managed through tools, but collaboration is fundamentally a human activity. Collaboration does not happen spontaneously, it needs to be actively designed and supported.

The quality of collaboration is difficult to assess objectively. It is often perceived as '*we know it when we see it*', and an assessor must frequently refer to proxies when seeking measurement, such as:

- Time taken to establish an agreed approach and to resolve issues
- Quality and value of the outcome(s) delivered
- Resources required to maintain effective performance (for example, to validate reports or establish whether evidence is consistent with claims)

Another term for this is 'trust'. The absence of trust creates friction in relationships and interactions. Trust means parties pay fairly and deliver on their promises. They can still argue robustly to establish what is fair, indeed, ignoring differences of opinion undermines trust and promotes abuse.

Defining what collaboration means, how it should be seen in action and how it will be measured is an important consideration at this stage. Collaboration should be:

- Active and energetic
- Focused on the delivery of value
- Flexible about means and approaches
- Forgiving and learning, quick to adapt when issues are discovered
- In for the long haul, not for immediate exploitation or the ability to win
- Equitable, with all parties willing to compromise

3.1.8.1 Coordination mechanisms

Coordination and collaboration may at first appear to be 'soft' and irrelevant. Should opposition be encountered, one useful technique is to construct a results chain,[22] a diagram that relates initiatives taken to the results they deliver in terms of outcomes for the SIAM ecosystem (for more on the results chain, see section **3.5.2.2 Develop a basic results chain**).

There are many mechanisms that can contribute to collaboration. The following list is provided for input into service provider selection and for inclusion in design of the SIAM model:

- Clear communication of the business vision and objectives
- Contractual obligations – for example, collaboration schedule, innovation schedule and fund
- Incentives – payment (direct, shared pool), additional business, executive access and governance recognition
- Process
- Tooling
- Informal networks
- Risk and issue resolution
- Structural elements and participation in these
- Onboarding workshops
- Face-to-face meetings
- Role and task definition
- Dispute resolution mechanisms and skills
- Operational level agreements (OLAs)/collaboration agreements (non-legal/contractual)
- Group events and activities
- Joint working parties/continual service improvement
- Vision, belief, leadership
- Organizational change management (OCM)
- Staff behavior guidance, feedback
- Mentoring and coaching
- Behavioral/cultural (notably leadership), training in skills related to cooperation
- Leadership selection – both task and relationship centered

Within the Plan & Build stage, the designers should consider these elements, although some may be deployed later as part of the Implement, or Run & Improve stages.

---

[22] *en.wikipedia.org/wiki/Benefits_realisation_management*. Results chain is also referred to as a benefits dependency network.

Incentives/penalties and interdependencies

The history of incentives and collaboration clauses in contracts has not been good. Many industry examples cite situations where contracts were robustly enforced by customers and subverted by service providers.

End-to-end success in service delivery is dependent upon the joint efforts of all the stakeholders involved. Therefore, there needs to be a degree of liability should any party fail to cooperate, and each service provider could have a proportion of its payment subject to the action of others.

For cooperation to work, all parties must play their part. This is a huge dependency that is often ignored. The customer organization is just as likely to be ill-suited to cooperation as are the service providers, with even more devastating effects.

Not all the mechanisms have to be introduced at the same time or built at the same rate. Consider selecting some for early evaluation, based on their contribution to the overall SIAM strategy. As in every strategic decision, concentrate your resources, make a choice!

3.1.8.2 Relationship management

Within a SIAM ecosystem, service providers' flexibility can help to deliver results. Service providers that refuse to engage with the SIAM model's process forums and working groups, or that insist, 'I've done what it says in the contract', can have a negative effect. Flexibility helps to develop the most effective collaborative working environment, which will often require changes to the way all layers within the SIAM model work.

The key features of relationship management in a SIAM context are:

- Shared vision and objectives
- Values and behaviors
- Creation of a collaborative environment:
  o Organization and governance
  o Risk management
  o Issues management and resolution
  o Knowledge management and sharing
  o Skill sets and training

Relationship measurement and improvement

An application service provider put in a bid for a SIAM-based contract that required 24/7 support for business-critical applications. As a part of the contract, it would be liable for a percentage of service credits (penalty) if its service line missed an end-to-end service target. This was even the case if the fault was found to be related to infrastructure, which was supported by the service provider that supported the wide area network (WAN).

The application provider had previously been working within non-SIAM contracts and queried during the bidding process why it should be penalized if the root cause was not related to the services for which it was contracted. The response provided, as part of clarification, was that service providers must be committed to end-to-end performance of the customer organization's business and services.

In response, the application service provider defined appropriate error logging and alerts at application level to notify potential issues in networking and to help to reduce the fault-finding time when working on incidents related to service.

3.1.8.3 Collaboration agreements

SIAM contract structures should include collaboration requirements, or be supplemented with collaboration agreements that codify the principles of collaboration and delivering services jointly for the benefit of the customer organization, rather than as disparate and siloed service providers. Collaboration agreements should be broad enough to apply to all service providers rather than a per provider agreement and should:

- Describe what is meant by collaboration
- Define what the objectives of the collaboration are
- Explain how the different parties will execute the collaboration
- Define any responsibilities regarding collaboration
- Outline how collaboration will be measured
- Define incentives in place to increase collaboration

Although it can be difficult to strictly enforce collaboration, these measures at least set the tone of the engagement and provide justification for raising the issue if collaboration is not forthcoming. Collaboration agreements are a new concept to many service providers unused to the workings of a SIAM ecosystem. At the heart of the challenge:

- A customer wants an integrated end-to-end service in which the service providers (internal and external) collaborate effectively
- A service provider wants a clear scope and autonomy to work and meet its own business objectives

Collaboration agreements need to serve all parties fairly and not put unnecessary limits on working practices, or hamper service providers by creating restrictions that will render their contribution untenable or unprofitable. This conflict requires a well-founded agreement. Customer organizations often opt for the two extremes: either no end-to-end obligations or an overly detailed collaboration framework. Neither of these is likely to be successful and therefore a negotiated approach, evolving over time, is often the most effective solution.

When seeking to construct an agreement in the presence of multiple service providers, the customer organization must recognize this dilemma. Service grouping can solve this apparent incompatibility (see section **2.5.5 Existing services and service groupings**), but requires a paradigm shift that allows the service integrator to be accountable for an end-to-end service, while being dependent on one or more service providers. Because SIAM assigns the accountability for an end-to-end service to the service integrator, a service provider can be certain of its individual scope and the customer organization can be assured that it will receive its desired outcomes. It is the service integrator's role to shield the customer organization from these 'boundary challenges'.

The recommended course of action is to:

1. Identify the outcomes and outputs sought from integration (see Discovery & Strategy).
2. Consider which party is responsible for each outcome or output, and the mechanism that is best placed to achieve them (for example, contracts are generally ineffective in governing behavior, so do not seek to contract for it. Contract for participation and direction by governance, and for the service integrator as managing agent).
3. Determine where standardization is to be applied (for example, selection of master toolset) and where discretion is available (for example, each service provider's internal tooling is acceptable if it integrates and exchanges with the master).
4. Develop the service agreement schedules considering both end to end and individual service provider accountabilities.
5. Develop the form for the contract and other governing artefacts, including integration and interface.

Much of the negotiation will concern the strength of end-to-end liability that is accepted by each service provider, and the price that is put upon it. Some service providers are inherently cooperative and will do so gladly. Some will resist strongly. If the customer organization chooses to select a resistant service provider as part of the ecosystem, issues may ensue. The cost and risk involved must be incorporated in the overall case.

Individual versus team selections

Some service providers are naturally cooperative, just as some athletes are natural team players. A team coach will select based on both individual talent and the ability to form a unified body that delivers results consistently on the field of play. The ability to form such a team to the satisfaction of team owners is notoriously difficult in sport, and coaches pay a high price for both success and failure.

It is strongly advised that factors associated with cooperative behavior are heavily weighted in the selection of prospective service providers, particularly for prospective service integrators. Some reluctant collaborators will enthusiastically welcome the concept of a collaboration agreement and then robustly oppose the imposition of any meaningful obligations. Any attempts at subversion should be highlighted and publicized for what they are. A customer organization wishing to work with collaborative service providers must be subject to the same principles and sign up for equivalent and material obligations.

Assessing collaborative behavior

Assessing potential service providers is more difficult in some markets than others. For example, in Switzerland, public organizations must follow World Trade Organisation (WTO) rules and issue official invitations to tender for services.

This very often hinders the commissioning organization from evaluating cooperative behavior, because only limited interaction is allowed during the bidding phase. This would mean not all the recommendations in this chapter would be possible.

3.1.8.4 Building cooperative teams

Research has suggested that creating teams that are large, virtual, diverse and composed of highly educated specialists is increasingly crucial within challenging projects. These same four characteristics make it hard for teams to get anything done.[23]

Collaboration between all layers is important and often omitted in the construction of a SIAM model. A SIAM model spans legal entities and seeks to maximize end-to-end benefits. In addition to skills and experience, of equal importance are the attitudes that foster collaboration in a SIAM ecosystem. These attitudes must be shared both across the SIAM layers between the customer organization's retained capabilities, the service integrator and the service providers, and laterally between service providers for the overall value chain to work seamlessly.

Some people are inherently cooperative, others are less so. A cliché widely and wisely applied is *'recruit for attitude; train for skills'*. Attitudes are far harder to instill than skills. It is not enough for a service provider, service integrator or customer organization to hope that collaborative attitudes will evolve between the parties over time.

The service integrator will need to be especially active in influencing service provider organizations. These elements can assist:

- The SIAM governance framework
- Cross-service provider forums
- Supplier management
- Performance management (with shared key performance indicators (KPIs))

Collaboration can occur without design but is likely to be confined to small areas and can take years to mature. A descent into silos, mutual recrimination, isolationism and distrust is far more likely, especially where teams are physically separated, and staff do not know each other.

There are eight factors that support successful team building,[24] and can be adapted to a SIAM ecosystem:

1. **Investing in signature relationship practices:** leaders can encourage collaborative behavior by making highly visible investments that demonstrate their commitment to collaboration. For example, in facilities with open floor plans to foster communication.
2. **Modelling collaborative behavior**: in companies where the senior executives demonstrate highly collaborative behavior themselves, teams collaborate well.
3. **Creating a 'sharing culture'**: mentoring and coaching, especially on an informal basis. Help people to build the networks they need to work across corporate boundaries.
4. **Ensuring the requisite skills**: HR departments that teach employees how to build relationships, communicate well and resolve conflicts creatively can have a major impact on team collaboration.

---

[23] *https://hbr.org/2007/11/eight-ways-to-build-collaborative-teams*.

[24] *https://hbr.org/2007/11/eight-ways-to-build-collaborative-teams*.

5. **Supporting a strong sense of community**: when people feel a sense of community, they are more comfortable reaching out to others and are more likely to share knowledge.

6. **Assigning team leaders that are both task and relationship oriented**: the debate has traditionally focused on whether a task or a relationship orientation creates better leadership, but in fact both are key to successfully leading a team. Typically, leaning more heavily on a task orientation at the outset of a project and shifting toward a relationship orientation once the work is underway works best.

7. **Building on heritage relationships**: when too many team members are strangers, people may be reluctant to share knowledge. A good practice is to put at least a few people who know one another on the team.

8. **Understanding role clarity and task ambiguity**: cooperation increases when the roles of individual team members are sharply defined, yet the team is given latitude on how to achieve the task.

Collaboration improves when the roles of individual team members are clearly defined and well understood, and when individuals feel that they can do a significant portion of their work independently. Without such clarity, team members are likely to waste too much energy negotiating roles or protecting their own position, rather than focusing on the task or required outcome.

3.1.8.5 Setting and managing shared goals

In the SIAM ecosystem, all stakeholders should be working towards shared goals related to a service(s) and performance. Shared goals are created based on the customer organization's requirements for the service. The service integrator (and potentially the service providers) may help to shape the goals based on the experience it brings.

Consider this in the context of a SIAM structure with an external service integrator. The customer organization will have clearly defined goals for the service(s) and should have communicated these to the service integrator as part of the procurement process. The service integrator then shares these goals with the service providers that need to integrate them into their own internal measures and targets.

However, all the organizations involved also have their own goals and targets. For example, the service integrator and one of the external service providers might be in competition in another area of the market. Either might want to make the other look ineffective to cause some reputational damage. They could attach more importance to this goal than to the shared goal related to the service.

Equally, an external service provider might hope to increase the number of service elements it is able to offer by making an internal service provider look weak. There is a lot of opportunity for individual organizational goals to come into conflict with shared goals in the SIAM ecosystem. The only way for a SIAM environment to be successful is for all parties involved in service provision to unite behind the shared goal. It must have more weight than any individual organizational goals, and any conflict must be identified and addressed.

Goals versus penalties

Immature SIAM ecosystems will document goals in contractual agreements and assume that takes care of everything. In reality, goals need to be based on shared values and relationships.

Communication and ongoing management of shared goals is essential for them to be effective. Remember, most customer organizations do not want to receive money back based on service penalties – they want the service that they ordered.

The service integrator is responsible for ensuring that goals are monitored and measured. Positive trends can be used to reinforce relationships and celebrate successes, and negative trends need to be analyzed to assess whether corrective action is needed. All the organizations involved in the SIAM ecosystem will have a role to play in measuring shared goals.

3.1.9 Tooling strategy

The tooling strategy should be created during the Plan & Build stage, under the governance of the solutions architecture and assurance functions. The technology required to provide the services and to support the processes is designed, including any necessary tools.

The current technology landscape should have been identified during the preceding Discovery & Strategy stage, and this information should be used to identify any gaps with the design, so that they can be addressed during the Implement stage. The tooling strategy should define policies for technology, for use by the service providers when they design the technology for their services.

Technology is relevant not only to services but also to a wide range of tools required to support the services. The choice of many of these tools is usually local to each organization, driven by corporate decisions that are unlikely to be changed by a transition to a SIAM model. However, each organization may need to integrate them with other parties in the ecosystem or use them in different ways.

This includes tools for:

- Service management processes
- Integration between different tools
- Integration between different services
- System and service monitoring
- End-to-end capacity and performance management
- End-to-end service reporting
- Collaboration
- Communication
- Resource management
- Records management
- Risk management
- Customer relationship management
- Financial management
- Order management
- Project/program management

This list is not exhaustive. The complete list depends on the type of organization and the chosen SIAM model, but it is important to consider the use of each tool within the SIAM ecosystem, especially the tools for which there is an element of choice or control. This is with regard to:

- Interactions with other parties
- The impact of technology obsolescence
- Security policies
- Availability requirements
- Other service level requirements
- Future change of provider
- Licensing
- Technology roadmaps
- Criticality within the SIAM ecosystem

3.1.9.1 Tooling strategy and roadmap

The outputs of the Discovery & Strategy stage will provide an input to the tooling strategy and roadmap. In the Plan & Build stage, it is important to review the initial intentions for tooling against the SIAM model, and the capabilities of the service integrator and service providers to create the tooling strategy, as decisions made at this point will be difficult and expensive to change later in the roadmap.

The tooling strategy should define:

- Technologies/tools in scope
- Mandated use of each toolset, as per the selected option
- Any technologies mandated
- Any technology policies
- Interoperability requirements
- Ownership of assets
- License management approach
- Integration approach
- Technology roadmap
- Release policy, to ensure that only supportable versions are used

It is necessary to establish the current baseline and capabilities of each service provider's toolsets. This is not possible in the Discovery & Strategy stage, as the service providers are not known. It is also crucial to establish the service providers' capability and willingness to make any toolset changes. There is little point in creating a tooling strategy to use a single toolset if the service providers have no desire to use it (and are not contractually obliged to).

In the Discovery & Strategy stage, the market for service integration tools should have been investigated to identify potential tools. The preferred tool should now be selected, considering the SIAM model and the capabilities of the service providers. There are four options described in the **SIAM Foundation BoK**:

1. A single toolset is used by all parties
2. The service providers use their own toolsets and integrate them with the service integrator's toolset
3. The service providers use their own toolsets and the service integrator integrates them with its own toolset

4. An integration service is used to integrate the toolsets of the service providers and the service integrator

It is crucial to make a well-informed decision on which of the four options for tooling is to be selected.

Table 5 shows the advantages and disadvantages of each option. The ideal scenario is option 1, as a single toolset means that there is only one data source. However, as described earlier in this publication, this is often not possible. If this option is not possible, the detailed design for the integration, data standards, interchange standards and associated policies must be carefully developed to ensure success.

Table 5: Toolset Options

Option	Advantages	Disadvantages
1. Single toolset	• Single source of data • No consolidation required for reporting	• Not suitable for commodity providers • Providers likely to have two toolsets, which they must then integrate locally • Maintaining licenses used by providers • Can increase bid costs • Can result in providers not bidding
2. Service providers use own toolset and integrate it with service integrator's toolset	• Providers continue to use existing toolset	• Not suitable for commodity providers • Can increase bid costs • Data needs consolidating for reporting • Some providers may be slow to integrate
3. Service providers use own toolset and service integrator integrates it with its toolset	• Providers continue to use existing toolset • Suitable for commodity providers	• Data needs consolidating for reporting
4. Use integration service	• Providers continue to use existing toolset • Suitable for commodity providers	• Data needs consolidating for reporting • Adds another service provider to the ecosystem

Option	Advantages	Disadvantages
	• Rapid implementation as preconfigured for commonly used toolsets and datasets • In-built audit trail • In-built status tracking	

These options can be combined, for example, using a single toolset for all except commodity service providers. This will enable the tooling strategy to contain:

- The current mode of operation current mode of operation (CMO), including the tools deployed, costs, license structures, support models and quality (for example, how well it supports the organization)
- The future mode of operation (FMO) in terms of each tool, whether the tool will be retained, retired or transferred to one of the service providers in the SIAM model
- A detailed cost analysis
- A functional model that describes which tools will support which processes and activities (such as monitoring, reporting)
- A transition plan that describes the approach to move from the current state to the future state
- How integration will transfer data between tools

This allows an understanding of the current capability so that gaps can be revealed and an understanding of ownership is clear. Understanding the tooling strategy and how the customer organization intends to leverage its providers from a technology innovation perspective is important. In some instances when transitioning to a SIAM model, a customer organization may decide it would be advantageous to upgrade to an industry standard tool that is used by most of the service provider organizations that they plan to work with.

In establishing these requirements, the customer organization needs to baseline the tools and technology in the CMO, and how and by which element or function within the organization this is utilized. One issue may be that in the FMO, one or more of the service providers may be providing and managing the current tool but may not be part of the future state plan.

3.1.9.2 Standard integration methods

In a SIAM ecosystem, it is common to experience data integration challenges, usually where different service providers use their own tools. Even if service providers adopt the customer organization or service integrator's own preferred tool, it is likely that they will still need to integrate with their own shared resources and tools.

Integration allows data to be shared seamlessly between different systems. **Table 6** shows various approaches to achieving this, each depending on the following factors:

- The volume of data to be shared
- The time criticality of the data
- The importance of the data

- The security implications of data being compromised
- The data segregation requirements
- Technology compatibility

Table 6: Various Data Integration Approaches

Approach	Advantages	Disadvantages
Manual data entry (including 'swivel chair' approach)	• Easy to implement	• Requires extra resource • Possible record duplication • High risk of loss of data integrity • High risk of erroneous data entry • Time delays
Manual bulk update	• Reduced risk of loss of data integrity	• Requires extra resources • Time delays
Automated bulk update	• No extra resources required	• Requires investment • Time delays
Automated data transmission	• Near real time • No extra resources required • Full data integrity	• Requires investment • Can take longer to implement

During the Plan & Build stage, the customer organization, with support from the service integrator, will develop integration approaches for each tool in the scope of the SIAM model. The integration approach adopted will vary depending upon the specific toolsets chosen or available, the data contained within it and the capabilities of the technology and the providers.

As part of an interim operating model, a 'swivel chair' approach is often used, which then evolves into full integration.

3.1.9.3 Ownership of data and toolsets

As part of the tooling strategy, the future ownership of data and toolsets must be established. When ownership is changing, any receiving party must formally accept that ownership has transferred to them. In addition to this transfer, training and data privacy implications and requirements must be considered, and suitable strategies developed to ensure that these requirements are met.

Another area to consider is the customer organization's desire to own certain datasets, such as locations, users, the ownership of process workflow scripts that enable automation, etc. It is also important to define identifiers in the common data dictionary, and to mandate use of these when exchanging data across the ecosystem, otherwise integration becomes highly complex. Examples of these identifiers are employee numbers to identify individual users, service identification codes/names and service provider codes/names.

Before the rise of SIAM models, organizations often outsourced all of these factors only to find that at the end of the sourcing arrangement they did not own any intellectual property (IP), data or automation flows. They needed to recreate datasets, processes and workflows from nothing. When defining the tooling strategy, it is a good time to consider and mitigate this risk (see section **2.5.5.3 Intellectual property**).

3.1.9.4 Ease of adding and removing service providers

The future state tooling strategy should provide the ability to easily onboard and offboard service providers. This needs to include the possibility of adding, changing or removing an external service integrator.

The aim is to allow service providers to be easily onboarded and offboarded, and reduce the dependency upon any service provider that may subsequently be offboarded (see sections **2.3.13.4 Onboarding and offboarding of service providers** and **3.3.3 Transition planning**).

3.1.9.5 Adopting a common data dictionary

In a SIAM ecosystem, a common data dictionary can reduce the potential complexity involved in bringing together data and systems from various service providers. The criticality and applicability of such a dictionary within a SIAM model, stems from the following principles:

- Service providers in a SIAM ecosystem will usually have their own processes and supporting glossary. For example, the definition of an incident versus a service request.
- There is a need to settle on a standard glossary.
- There is a need to adopt a common model for logging and tracking all items in the SIAM model in a common way. For example, the statuses assigned to an incident record as it passes through the workflow (opened, assigned, resolved, on-hold, closed, etc.).
- There is a need to settle on a single master data source for base data, such as services, locations, users, etc., and to ensure that these are kept in sync with the various tools and service providers working within the SIAM ecosystem.
- There is a need to own and track the data model and protect the integrity of the data model through change control methods.

Essentially, a common data dictionary can be a central repository or tool that enables the communication of stakeholder requirements in such a way that the customer organization, service integrator and service providers can configure their tools to meet those requirements more easily. In its truest form, a common data dictionary provides detailed information about the business or organization's data, and outlines the standard definitions of data elements, their meanings and allowable values.

From a conceptual or logical perspective, some organizations use an entity relationship diagram (ERD) to focus on the high-level business concepts. The common data dictionary provides a more granular and detailed view about business concept attributes.

Key elements of a common data dictionary

A common data dictionary provides information about each attribute. This information may also be referred to as fields in a data model, where an attribute is placed in a database that holds information.

SIAM governance article

If a common data dictionary was created for an article on SIAM governance, the attributes might be:

- Title
- Author
- Category
- Content

A common data dictionary is often organized as a spreadsheet. Each attribute is listed as a row in the spreadsheet, and each column labels an element of information that is useful to know about the attribute.

The most common elements included in a data dictionary are:

- **Attribute name:** a unique identifier, typically expressed in business language, which labels each attribute
- **Optional/required:** indicates whether information is required in an attribute before a record can be saved
- **Attribute type:** defines what type of data is allowable in a field. Common types include text, numeric, date/time, enumerated list, look-ups, Boolean and unique identifiers

Although these are the core elements of a common data dictionary, it is not uncommon to document additional information about each element. This may include the source of the information, the table or concept in which the attribute is contained, the physical database field name, the field length and any default values, etc.

3.1.10 Ongoing improvement framework

During the Plan & Build stage, an improvement framework is developed and maintained in conjunction with all parties in place within the SIAM model at that point. The framework will ensure a focus on continual improvement across the SIAM ecosystem. Service providers should have incentives that encourage them to suggest and deliver improvements and innovation through their people, processes and technology.

All processes should be subject to review and improvement measures where possible. This continual improvement can be managed within the area responsible for the provision and fulfilment of the process, but should also feed into the larger continual service improvement process and related process forum, particularly when an improvement is dependent on sources external to the process or introduces significant risk or cost.

Theory of Constraints

The Theory of Constraints (ToC[25]) is a methodology for identifying the most important limiting factor (constraint) that stands in the way of achieving a goal, and then systematically improving that constraint until it is no longer the limiting factor. In manufacturing, the constraint is often referred to as a bottleneck. This methodology is useful when designing the SIAM model, and in the planning and execution of improvements.

ToC takes a scientific approach to improvement. It hypothesizes that every complex system consists of multiple linked activities, one of which acts as a constraint upon the entire system (the constraint activity is the 'weakest link in the chain'). To function more efficiently, organizations need to identify constraints and reduce the impact of the bottleneck or remove it completely. Not all constraints can be eliminated, but their effects can be reduced. For example, a regulatory requirement might be a constraint.

ToC recognizes that:

- Every process has at least one constraint that can affect its ability to meet its goal
- A process is only as successful as its weakest link and can only operate to the capacity of its constraints
- Improving constraints is the fastest and most efficient way to improve the entire process or system

ToC includes a sophisticated problem-solving methodology called the 'thinking processes'. The thinking processes are optimized for complex systems with many interdependencies (for example, service lines). They are designed as scientific 'cause and effect' tools that strive to first identify the root causes of undesirable effects (UDEs), and then remove UDEs without creating new ones.

The thinking processes are used to answer the following three questions that are essential to the ToC:

- What needs to be changed?
- What should it be changed to?
- What actions will cause the change?

Table 7 shows the five focusing steps in ToC.

---

[25] *www.tocinstitute.org/theory-of-constraints.html*.

Table 7: Five Focusing Steps

Step	Objective
Identify	Identify the current constraint (the single part of the process that limits the rate at which the goal is achieved).
Exploit	Make quick improvements to the throughput of the constraint using existing resources (make the most of what you have).
Subordinate	Review all other activities in the process to ensure that they are aligned with and truly support the needs of the constraint.
Elevate	If the constraint still exists and has not moved, consider which further actions can be taken to eliminate it from being the constraint. Normally, actions are continued at this step until the constraint has been 'broken' (until it has moved somewhere else). In some cases, capital investment may be required.
Repeat	The five focusing steps are a continual improvement cycle. Once a constraint is resolved the next constraint should immediately be addressed. This step is a reminder to never become complacent – aggressively improve the current constraint … and then immediately move on to the next constraint.

3.2 Organizational change management approach

When an organization makes a strategic decision to embark on a SIAM transition, effective organizational change management (OCM) is essential. Changing to a SIAM environment is a major change, so OCM must start as soon as possible and must be in progress **before** the Implement stage. Organizations that leave OCM until the last minute, or do not give it the necessary attention, are unlikely to be successful with a move to a SIAM model.

Although a SIAM transition project involves change at many levels, including processes, job roles, organizational structures and technology, it is the individuals (from all SIAM layers) who must ultimately change how they work. If these individuals are unsuccessful in their personal transition and do not embrace and learn a new way of working, the overall transition will fail.

OCM is the discipline that helps to prepare, equip and support individuals to successfully adopt change. There is considerable research demonstrating that a structured approach for supporting individuals in an organization will help increase the likelihood of change success. There are several industry-developed approaches available to help support organizational change, such as ADKAR®, alternatively, an organization may have evolved its own proprietary approach.

Whatever approach is taken should be goal-oriented, allowing agents of change to focus their activities on what will drive individual change and therefore allow for organizational results. Clear goals and outcomes for change management activities need to be set by the customer organization and delivered by the service integrator.

Models such as ADKAR outline the individual's successful journey through change. Each step of the model naturally fits into the typical activities associated with change management. In identifying the required goals or outcomes of the change, the approach or model chosen should provide a framework for change management teams in both the planning and the execution of their work.

ADKAR

ADKAR is a research-based, individual change model that represents the five milestones an individual must achieve to change successfully. The model can be extended to show how businesses can increase the likelihood that their changes are implemented successfully. The ADKAR acronym represents the five milestones an individual must achieve for change to be successful:

Awareness

Represents a person's understanding of the nature of the change. Awareness also includes information concerning the internal and external drivers that created the need for change, as well as the WIIFM, or 'What's in it for me?'. Awareness is the outcome of an individual's understanding of the nature of the change, why it is needed and the risk associated with not changing.

Desire

Represents the ultimate choice of an individual to support and engage in a change. Desire to change remains elusive, since it is driven by an individual's personal situation as well as intrinsic motivators unique to each person. Success comes when four factors are understood and catered for:

- Nature of the change and the WIIFM
- Organization or environmental context
- An individual's personal situation
- Intrinsic motivation

A combination of these factors will ultimately contribute to the behaviors expressed by an individual when confronted with change. Desire is the outcome of stakeholder management and resistance management.

Knowledge

Represents how to implement change. When a person has the Awareness of a need to change and the Desire to participate in the change, Knowledge is the next building block for realizing that change. Knowledge includes training and education on the skills, behaviors, processes, systems, and roles and responsibilities required. Knowledge acquisition is the outcome of training and coaching.

Ability

Represents the demonstrated capability to implement the change and achieve the desired performance level. The presence of knowledge is often insufficient in isolation. Training and coaching does not equal practical capability and competency. Additionally, some individuals may never develop the required capabilities. Ability is the goal or outcome of additional coaching, practice and time.

Reinforcement

Is to ensure change becomes embedded. Reinforcement is required to prevent reversion back to old habits, processes or behavior. This stage ensures elements are present to mitigate this. It seeks to address the incentives or consequences for not continuing to act in the new way and re-visit knowledge and ability milestones. Reinforcement is the outcome of adoption measurement, corrective actions and recognition of successful change.

The goals or outcomes defined by ADKAR are sequential and cumulative (they must be achieved in order). For a change to be implemented and sustained, an individual must progress through each of the milestones, starting with Awareness.

3.2.1 Organizational change in a SIAM ecosystem

For a SIAM transformation, organizational change management (OCM) commences soon after approval. It forms part of the Plan & Build stage, where the SIAM model and the OCM approach are defined. OCM's main activities then occur within the Implement stage.

The agreed approach to OCM should provide a sequence for managing the people or individual aspects of change. The people are all the stakeholders involved in the customer organization, the service integrator and the service providers. Organizationally, the approach provides a foundation for change activity, including sponsorship, readiness assessments, communications, coaching, training, and stakeholder and resistance management.

In multi-sourced service delivery models, the key to success is the ability to manage the challenge of cross-functional, cross-process, cross-service provider integration while finding an effective method for controlling this delivery environment. The boundaries of responsibility for the OCM approach must be clear and consider accountability for:

- The overall OCM program
- The OCM of the service integrator
- OCM within each service provider
- OCM related to customers and other stakeholders

The method and approach for each of these may not be the same. It would be unlikely, for example, to have a service integrator responsible for OCM within a service provider. This may be best done together to get a common experience and understanding.

The overall OCM program will set the expectations, provide an environment to encourage good behavior and measure the outcome, but it does not have management control of the service providers' staff. Each service provider should have a contact responsible for its OCM working with the overall program.

3.2.2 Challenges for organizational change activity

There are several organizational change challenges that are specific to a SIAM ecosystem.

3.2.2.1 Integrating new service providers

One aim of SIAM is to provide the customer with the opportunity to select service providers in line with their changing needs and overall strategy. OCM can provide support when service providers are added or removed. These events affect working relationships and interfaces, and can amount to considerable change activity.

3.2.2.2 Lack of clarity over governance

SIAM provides governance over service providers on behalf of the customer organization, while giving them the freedom to fulfil their role against an understood mandate. Some of the service providers will be external and typically answer to their own shareholders and have their own commercial interests at play. These interests could conflict with those of the customer organization and/or other service providers. OCM can be used to create understanding of the changing structure and responsibilities required.

3.2.2.3 Changes to working practice

SIAM establishes the necessary outcome-focused capability to extract value from multiple service providers. Part of the service integrator role is to provide governance for services according to established standards and policies of the customer organization. For many service providers, and in some cases for the customer organization's retained capabilities, this service integration role is a change to existing working practices.

For example, in some situations, service providers are required to adopt the customer organization's technology solution, which means a transition to new tools. For many service providers, the tools they use are embedded in their approach and ways of working. However, if the customer organization wishes to integrate tooling across different service providers, OCM can support building the required knowledge and ability to support this change.

A critical success factor (CSF) for SIAM is demonstrating the impartiality of the service integrator, as well as its ability to leverage control over all the service providers. Internal service providers may resent control by an external organization, and external service providers may revert to working to contractual directives and effectively 'painting by numbers' within the defined constraints of a contract.

Taking the five elements that define the basic building blocks for successful change awareness, desire, knowledge, ability and reinforcement (ADKAR) and using them as principles, OCM can be used:

- As a coaching tool to support individuals through process or changes to team dynamics
- To guide activities such as communications, sponsorship, coaching and training
- To diagnose struggling functions or processes

When OCM is used successfully, service providers are more engaged. Organizational working structures and process collaboration are adopted faster. Individuals contribute ideas and suggest new ways of working. Peer-to-peer working groups across service providers are in operation. Flexibility and adaptability become parts of the value system, essential in a model that incorporates the concepts of separation of concerns, modularity and loose coupling.

All of these elements provide the flexibility SIAM needs to be able to decouple underperforming service providers. The ADKAR model provides a goal-oriented framework to support change to a service provision model, in this case based on SIAM.

Missing ADKAR elements

In the workplace, missing or weak elements of the ADKAR model can undermine business activities. In the absence of Awareness and Desire, it is likely there will be more resistance from provider teams leading to slower adoption of working practices, staff turnover and delays in optimizing the SIAM model. If Awareness and Desire are low, failure is likely.

In the absence of Knowledge and Ability, there will often be less cohesion between working groups or committees, a fundamental structural element for SIAM success.

In the absence of Reinforcement, service providers may lose interest and revert to old working practices and approaches. Such consequences will ultimately affect the success of the SIAM model negatively.

3.2.2.4 Communicating change information

The right people need the right information at the right time. Within a SIAM ecosystem, consideration must be given to the design of communication activities, including aspects such as medium, frequency and message.

Effective messaging benefits from a wide range of skills and expertise in the participant pool. SIAM ecosystems provide the opportunity to engage with a diverse group of individuals, and it is important to grasp the opportunities. Structural elements in the form of boards, working groups and process forums provide an excellent medium for engagement and message transfer.

3.2.3 Virtual and cross-functional teams

To succeed in the global economy, organizations frequently rely on a geographically dispersed workforce. The success of a SIAM model involves building teams that offer the best functional expertise, combined with deep, local knowledge of specific markets and industries. Drawing on the benefits of service provider diversity, bringing together people with varied work experiences and different perspectives on challenges, helps organizations to obtain value from the SIAM ecosystem.

Cross-functional team definition

"A group of people with different functional expertise working towards a common goal. It may include people from finance, marketing, operations and HR departments. Typically, it includes employees from all levels of an organization."[26]

Managers who lead global teams face many challenges. Creating successful working groups is difficult enough when everyone is geographically co-located, and people share the same environment or work within the same organization. When team members come from different countries and backgrounds, are working in different locations for different organizations using different tools, communication can

---

[26] Wikipedia, *https://en.wikipedia.org/wiki/Cross-functional_team*.

deteriorate rapidly. Misunderstanding can ensue, and cooperation can degenerate into distrust. Distrust is the enemy of a successful SIAM ecosystem, where mutual respect and cohesion should be the core.

Staff within a SIAM ecosystem, no matter where they are located, must consider themselves to be colleagues or partners of all other ecosystem staff, as if the entire SIAM ecosystem were working in a single location and organization. It is important to establish good practices for creating a community atmosphere across service provider teams.

Getting cross provider teams to work together

While establishing the working governance and internal agreements for a SIAM model within a global Irish bank, a meeting was initiated with the four service providers. In the meeting room, the team sat at individual tables, company by company, which was not conducive to collaboration, especially when they were about to talk about how to work together.

The consultant who was responsible for delivering the new SIAM model and defining this collaborative working, decided that as an exercise he would mix up individuals from the service providers and sit them at different tables. He restructured the room so that at each table there were individuals from each organization. He then set them a task of writing a funny story from a scenario he provided and offered a prize for the best.

Once this was completed and everyone had relaxed and shared their stories, they assumed they would move back to their appropriate tables (with their colleagues). The consultant advised them to stay seated with their new colleagues and work in the newly formed teams to start discussing the main topic, 'Working governance and OLAs'. This time as a collaborative team.

Work practices (such as the definition of quality, acceptable incident response times or KPI calculation methods) should be standardized or, where not standardized, differences must be documented, accepted and published as working standards.

The following considerations are important:

- Identify appropriate behaviors
- Identify possible issues associated with evident behaviors
- Define value/belief expectations and explanations (describe global values and beliefs)
- Describe effective ways of interacting

Consider the following dimensions:

- Alignment to SIAM vision
- Profound understanding of what differentiates the service integration layer from traditional service provider activities
- Shared value statement examples
- How do we share information?
- How do we communicate?
- How do we make decisions?
- What is the role of the service integrator?
- How to deal with different histories and work styles of each service provider

The degree of emotional connection among team members is important. When people in a team all work for the same organization, the level of 'social distance' is usually low. Even if they come from different backgrounds, people can interact formally and informally and build trust. This will allow them to arrive at a common understanding of what certain behaviors mean, what is acceptable and what is not. If team members feel close and congenial, this will foster effective teamwork.

Co-workers who are separated geographically or are from different organizations may find it more difficult to connect and align. They may experience high levels of social distance and often find it difficult to develop effective interactions. Mitigating social distance is a management challenge for the service integrator.

In a remote team, culture-building activities are harder to implement, although not impossible. There are some differences between remote and co-located teams that must be considered, such as not being able to meet physically, time zone differences or shared working spaces (perhaps restricting noise, providing security, space). Technology issues may also impact collaboration options.

It is the duty of the service integrator to develop cohesion and low levels of social distance by:

- Being transparent – transparency is important for all teams, but it is even more useful for building culture in a remote team – be transparent about who does what (for instance, using collaboration tools)
- Enabling 'virtual water cooler' chats
- Creating an online collaboration/messaging space open for the teams to chat with one another and discuss non work-related topics
- Meeting in person when possible – everything from attending a pub quiz or taking part in a sporting event – being careful to not only focus on one type of activity that excludes people (for example, those who do not drink alcohol or those who have a disability)
- Organizing 'team challenges'

Whoever wants to participate should be encouraged to join in, but do not force anyone to do so.

Social events

In a new SIAM ecosystem, service providers were reluctant to engage with each other. The service integrator set up process forums to deal with specific areas of concern, and to get individuals from the service providers to work together, but these had only limited success.

The service integrator then organized a golf tournament, followed by a meal. Each team was made up of individuals from the different service providers and the service integrator. The event was a great success. Once taken out of the work environment, individuals found that they shared interests and former colleagues with one another, and were now more likely to engage with one another on work-related issues.

3.2.3.1 Challenges related to cross-functional teams

The common challenges that arise in cross-functional teams are:

- **Conflicting objectives**: each stakeholder is likely to have slightly different interests, which may lead to conflicting objectives
- **Reluctance to share**: it is a common challenge that there is some reluctance to share information, especially between competing organizations
- **Lack of automation**: this can hinder the ability to communicate or collaborate across teams, and may lead to frustration, double handling, and wasted resources and time

---

Cross-functional team examples

A team working on a major incident where the root cause is unclear often includes staff from the customer organization, the service integrator and multiple service providers. They will need to work together towards a shared goal (incident resolution), while meeting service requirements and balancing their own conflicting goals.

Similarly, a change management or change advisory board, involving staff from the service integrator and multiple service providers, would need to work together to review, prioritize and risk assess changes to a service to ensure beneficial changes were made successfully.

3.2.3.2 Creating shared goals

Shared goals allow people from disparate teams, functions or organizations to work together towards a single outcome. Goals might be quantitative and performance based (for example, priority 2 incidents will be resolved within four hours), or they might be more qualitative and related to culture and working practices (such as, we will operate no-blame post-incident reviews).

Once the goals are agreed at a senior level by the customer organization, they then need to flow down the layers of the SIAM model. If a service target relies on all incidents being resolved within four hours, the service desk technician needs to understand that as well as the senior manager.

Goal definition

Many organizations start from the assumption that defining goals is easy. It is not! Poorly defined goals based on ambiguous language and outcomes can drive poor behavior and damage relationships between the organizations in the SIAM ecosystem.

Time spent getting goals right delivers real benefit. Correcting things after contracts have been signed is more time consuming and expensive.

It is an important part of the feedback loop to celebrate when goals are achieved, this encourages the desired behavior and helps cement the group identity.

3.2.3.3 Collaborative practice

Culture is an important part of establishing an effective approach to collaboration. All workplaces have a culture shaping how things get done, how people interact and engage with others, and what is appropriate and acceptable behavior. This directly influences the management, organization and application of work. Culture is defined by the collective view of all parts of an organization. The challenge in a SIAM model is that the customer organization is only one layer of the overall ecosystem. Building the SIAM ecosystem is essentially about building a culture.

The role of the service integrator is to establish a unified and positive culture across all layers, structural elements and functions. Without a focused effort to introduce and maintain culture, it will be impossible to enable consistent and valuable activity, interactions and relationships.

Customer experience visits

A service integrator in the health service identified that its own staff and the staff in the service providers did not seem to be aware of the impact of service downtime on the users. Even though service levels were being achieved, satisfaction was low.

It decided to start a customer experience program for all existing and any new staff. In consultation with the customer, it designed a program so that the staff could visit and meet with users in their work environment, hospitals and doctors' practices, to understand just how they used the services, and what it was like when the services weren't available.

The customer experience visits had a profound and long-lasting effect and resulted in sustained improvements in attitude, behavior and culture. Some staff were visibly upset when they realized the effect of downtime. The visits helped everyone involved to focus on the outcomes, and not on the contracts.

A cohesive culture will enable the creation of an effective SIAM ecosystem where beneficial stakeholder and work outcomes generate positive thinking, while a poor culture will have a negative impact on both stakeholder and business performance. Culture is not only necessary for day-to-day operations, it is also important in attracting and retaining skilled staff and service providers.

3.2.3.4 The importance of good leadership

Although everyone has a role in creating a positive working environment, a positive workplace culture starts from the top, with the leadership team within the customer layer. The words and actions of leaders here and within the leadership teams of the service providers and service integrator demonstrate examples for others to follow.

The service integrator is accountable for the development of a positive environment and must:

- Give clear and consistent instructions and messages that reinforce positive behavior
- Demonstrate the type of behavior they expect from others
- Deal promptly with issues that may undermine the ecosystem
- Recognize and celebrate individual employee and service provider successes

The evident culture within a SIAM ecosystem is created from the collective view of all the individuals working there. It follows that getting the right mix and fit of people is important, either through selection or organizational change management (OCM) initiatives. People have different personalities, views and values, and understanding and recognizing these differences is essential to creating a positive work environment. The right workplace 'fit' is where the values and behaviors of an individual align with those of the people they work with and those of the customer.

In organizations where there are high levels of accord and a positive outlook, individuals will generally experience:

- Greater job satisfaction
- Closer alignment to the business (customer organization)
- Stronger loyalty
- Stronger commitment
- Lower absenteeism
- Higher levels of job performance

Individuals working in roles they like and for businesses that share their personal values, will typically be happier and more productive, as they feel part of the business and its outcomes. This is essential for SIAM success.

3.2.4 Conflict resolution

When a dispute surfaces and conflict resolution is required, the outcome can be predictable: the conflict escalates, with each side blaming the other in increasingly strident terms. The dispute may end in litigation, and the relationship may be forever damaged (see also sections **3.1.3.7 Incentives** and **3.1.3.9 Dispute management**).

The role of the service integrator is to ensure structural elements, in particular process forums, are available to allow early conversations before situations become irrevocable and to be the source of conflict resolution when required.

There are many potential sources of conflict in both personal and professional life, including:

- Different ideas and goals
- Unclear goals (conflict occurs because they have been interpreted in different ways)
- Poor communication
- Clashing personalities and types
- Different ways of working
- People or teams feeling that their needs are not being met
- Legacy issues potentially impacting future encounters
- People deliberately causing conflict because of their own agenda (which might include getting power or status, or punishing another person or team)
- Resistance to change, which can be deliberate or unintentional
- Weak/poor management and leadership

A SIAM model can amplify these because of the number of organizations involved. Different organizations and individuals will have different goals and ideas that must somehow be unified.

The following three negotiation strategies[27] form useful tools for the service integrator, the service providers and customer-retained capabilities when helping to resolve conflict, repair relationships, avoid the complications of legal proceedings and even create value:

1. Avoid emotional responses
2. Do not abandon value-creating strategies
3. Use time advantageously

Avoid emotional responses

Within the 'power play' that can be a characteristic of a multi-service provider model, service providers may make moves to assert their power. If the service integrator is not effective, it may struggle to maintain its authority and demonstrate impartiality. Discussions about seemingly rational issues can lead to angry outbursts, hurt feelings and simmering conflict.

By challenging and demeaning the other party (whether consciously or not), a service provider may be attempting to provoke an emotional response that will shift the balance of power in its favor.

Power play counter move

Consider techniques such as naming the move, that is, letting a service provider know that you recognize it as a power play. Alternatively, divert the move by shifting the focus back to the issue at hand.

Do not abandon value-creating strategies

Service integrators that understand the importance of collaborating with service providers to create value, may nonetheless abandon that approach during dispute resolution. Treating disputes as different from other aspects of deal making, they may view business dispute resolution as one in which only a single issue (such as money) is at stake. Consequently, they tend to look at the dispute resolution process as a win-lose battle, to their detriment.

In this situation, it is important to capitalize on shared interests, or non-competitive similarities.

Capitalize on shared interests

For example, perhaps both parties would suffer reputational damage if their dispute went public. In this instance, they might agree to keep certain aspects of their dispute resolution process confidential.

Reaching agreement on seemingly peripheral issues can help parties build a foundation of trust and optimism that enables them to resolve the main sources of their conflict. Disputants may also be able to create value by trading on their differing preferences and priorities.

---

[27] *www.pon.harvard.edu/daily/dispute-resolution/3-negotiation-strategies-for-conflict-resolution/*.

Trade-offs

If service provider A places a high value on receiving a formal apology from service provider B, service provider B might be willing to grant the apology in exchange for a lower settlement payment to service provider A.

Through such trade-offs, negotiators can increase the odds of a peaceful and lasting resolution.

Use time advantageously

The perceptions held about a dispute resolution process may change over time owing to experiences when dealing with the conflict and with the other parties involved. Rescheduling meetings, taking a few points of contention at a time and remaining in contact with the other parties during dispute resolution are useful approaches for the service integrator to employ.

Doing so may encourage the parties involved to see that existing approaches to the conflict are not working and that the prospect of negotiation offers some hope of improvement. When parties recognize the importance of meeting regularly, they may be able to work through their differences slowly.

3.3 Planning the approaches for onboarding service

The Plan & Build stage is focused on providing the design for the SIAM model and creating plans for the transition. Ideally, the service integrator should be selected and in place before the SIAM model is fully finalized and any service providers are selected.

If this can be achieved, the service integrator can be involved with the Plan & Build activities. The benefits of this approach are:

- The service integrator is involved with the design of service groups and selection of service providers, so it can use its experience to assist with these activities. In this instance, impartiality must be demonstrated by the service integrator. A good approach is to have a different team of resources engaged in this consulting capacity versus those who undertake the day-to-day integration activities.
- The service integrator is fully aware of the requirements placed on the service providers during selection and appointment.

These are ideal states that apply in the absence of constraints and preconditions. In reality, many organizations appoint the service integrator and the service providers at the same time, and the service integrator must get on with the job with the information it is given. There may be restrictions that prevent organizations involved in the Plan & Build stage taking a role in the SIAM ecosystem, because of perceived partiality issues.

Conflict of interests

A public-sector customer organization in Europe engaged an organization to help define its SIAM model. When it set out to procure its external providers, the same organization won the contract for the service integrator.

> The procurement was challenged by one of the unsuccessful bidders under the European procurement rules for public-sector organizations, claiming that the appointed service integrator had an unfair advantage. The unsuccessful bidder won its claim.

'Greenfield' environments, with no existing services or service providers, are rarely found in IT operations. There are normally services in place that need to be accommodated. The ideal scenario in a greenfield environment would be to:

- Define governance and structure as part of the Discovery & Strategy stage.
- Define what the service integrator and each service provider does and the outline of how to do it. The outline plan will be defined as part of the Discovery & Strategy stage, and more comprehensive plans will be created during the Plan & Build stage.
- Appoint the service integrator.
- Build and refine the SIAM model and coordination mechanisms.
- Appoint, onboard and induct the service providers (starting with the most integrated, progressing to the least).
- Refine and adapt the operational model and performance in the light of experience and developing business need.

The rationale for this approach is:

- Governance is at the heart of the Discovery & Strategy stage and must therefore be present to at least a minimal degree from the start. This guides and controls the other roadmap stages.
- The service integrator can assist with the design of the detailed SIAM model and the selection of the service providers, leveraging their experience.
- The service integrator is the most interconnected function, so early appointment would allow the SIAM model to be designed with integration at the heart.
- The service providers do not exist in isolation (for example, application services depend on infrastructure). If the most interconnected are established first, it is easier to add in the simpler service providers later, with the least overall change cost.

Because of the degree of interconnectedness present in multi-service provider operations and the rapid rate of business change commonly encountered, it is impossible to predict with certainty the best overall design. If such a state were to be developed, it could soon be out of date. The aspiration for operations is therefore one of dynamic stability and evolution.

In the real world, there are:

- Incumbent service providers with contracts that have time to run, interests to serve and costs incurred to effect change.
- Uncertain interactions and interfaces, some documented – some not.
- Changing business requirements, regulation, competitive changes and service provider changes.
- Change costs related to the formation and change of agreements, operational implementation, tooling and integration.

Moving to a SIAM ecosystem can be a complex endeavor. Many organizations choose to implement their model in a staged or phased way (see section **4.1 Comparing implementation approaches**), but the initial plans may need to consider several options.

A customer organization may start with an approach that it would like to see implemented. If the incumbents cannot be persuaded to accept the proposed change, the planned approach will need to be revised.

The service provider puzzle

Constructing a coherent service provider ecosystem may be compared to constructing a multi-dimensional puzzle. The requirement and interactions of each piece are defined by the space that piece will occupy. It is difficult to see what that space is before its neighbors are in place.

The role of the service integrator can be compared to that of the puzzle constructor. If there is a guiding brain that calls for pieces of a certain size, configuration and interface, there is a greater hope that the components may fit than if the approach is to go to market, calling aloud for a random bag of pieces from service providers that have not designed their services with each other in mind.

The construction of such an ecosystem needs to start with establishing how the pieces are to work together, building the mechanism into the service transition and maturing its practice over time. This rationale is one of the strongest behind the injunction to start with the service integrator layer and then progress to the service provider layer.

Note that if the service integrator and service providers are appointed at the same time, this can be like giving the service integrator the puzzle pieces but not the whole picture.

3.3.1 Onboarding process

Onboarding is the process of bringing new service providers into the SIAM ecosystem. This begins after appointment and usually continues until the service providers have 'normalized' their service (in the Run & Improve stage). As indicated previously, for successful onboarding each service provider or team must acquire the necessary knowledge, understanding and tools. Defined systems and process outcomes provide for consistent experiences – and where there is consistency, there is an ability to track effectiveness, efficiency and quality.

Onboarding allows the service integrator to welcome, set expectations and educate service providers so that all parties can work together effectively. It is also about providing and receiving information, and setting the tone for successful relationships.

The formality and comprehensiveness of an onboarding program varies widely across organizations. Some organizations prefer a more structured and systematic approach, while others follow a 'sink or swim' approach in which new service providers may struggle to figure out what is expected of them and the existing norms of the SIAM model.

Onboarding can vary in many ways, including in its formality, sequencing, number of onboarded service providers and how supportive the process is of new service providers. Having a clear onboarding process can facilitate the transition, increase confidence and satisfaction, improve productivity and performance, and reduce stress and confusion.

Some of the tactics used in this process include meetings, hangouts, social events, calls, videos as well as printed materials, digital communication, training documentation, tasks and reminders, forms, questionnaires and research.

The following approach outlines organizational practices for onboarding. Articulating business goals is a primary concern so the customer organization and its business goals are understood. It is great to have technology, tools and processes in place, but understanding business goals can ensure that the onboarding process is in line with what really matters. Lead with business goals and processes should follow naturally.

Organizational practices for onboarding include:

- Create a standardized onboarding process
- Define a reporting framework for management information, SLAs and KPIs (see section **3.1.7.2 Develop a performance measurement plan**)
- Design and use standardized acceptance criteria for each process area and practice
- Use risk management to reduce risks
- Prioritize process areas that need to be in place for go-live over ones that can be introduced later
- Design and implement formal orientation programs
- Create and use written onboarding plans
- Be participatory
- Use technology to facilitate the process
- Monitor progress over time
- Use milestones, such as 30 – 60 – 90 – 120 days, up to one-year post-organizational entry
- Obtain and act on feedback from service providers

A successful onboarding process maps all aspects of service delivery to the customer organization's business goals. Effective onboarding ensures that new service providers feel welcome and prepared, in turn giving them the confidence and helping them deploy the correct level of resources to make a positive impact within the SIAM model.

These areas should be considered when onboarding service providers:

- Reducing impact to end users
- Minimizing impact to the customer's business where possible
- Always having a target end state in mind
- Minimizing changes to toolset(s), and assessing time and cost related to manual versus automated operations
- Minimizing impact to existing services
- Ensuring clear accountability and ownership

3.3.1.1 Sequencing the appointment of service providers

The sequence by which service providers are engaged can have a significant impact on the operational effectiveness of the SIAM model and upon the costs of transitioning to and maintaining the model. An approach that starts with strategic service providers, then deals with tactical ones and then operational and commodity service providers is generally adopted to minimize the risks and establish high-priority services first.

However, there are often other practicalities and constraints to be considered, including:

- Based on the 80/20 rule, 20% of service providers will typically account for 80% of organizational spend. By focusing on that 20%, the customer organization will get more benefit from the transition, even while onboarding a smaller number of providers. Then, once these service providers are in place, the focus can be expanded to the next tier.
- The focus could initially be on the service providers that have the technology in place to connect to existing organizational tools.
- Onboard those service providers with least associated operational change first – for example, incumbent service providers may just have the same people doing the same job, meaning the end users are less affected by the change.
- If the most interconnected are established first, it is easier to add in the simpler service providers later.

Given the nature of a SIAM transition with so many changes happening in a small timeframe, it is important to carefully consider the impact on end users and limit disruption where possible.

Onboarding sequence example

An example of sequencing could be data center, remote hosting, voice, managed print, LAN/WAN and security (protective monitoring) service providers first, as they will set the infrastructure in place for other service providers. This could then be followed by end-user computing and other services, including application services.

3.3.1.2 Risks and issues during the appointment of service providers

First impressions have a lasting effect, including for the onboarding program. Poor onboarding and induction programs create risks, including low productivity, poor customer service, compliance risks, dissatisfied customers and users, and high staff turnover. The service integrator must seek to reduce these risks.

To avoid confusion and minimize risks, the roles of the service integrator and the customer organization in appointing and onboarding service providers must be very clear. If the advice in this publication on the stages of the SIAM roadmap are not followed, there is a risk that a customer may commission incompatible services and then ask a service integrator to put them together and make them work. Following the advice in this publication can help to avoid this situation (see section **3.1.2 Sourcing approach and the selected SIAM structure**).

Poor collaboration from existing service providers involved in onboarding new service providers can be a significant risk that also needs to be managed.

3.3.1.3 Options for onboarding

The service integrator role includes onboarding new service providers. This normally involves the transfer of policy, process and standards for documentation, and is often supported by scheduling workshops to introduce new approaches and working models, and resolve operational details of implementation. The customer-retained capabilities should also be involved, mainly around the more formal activities such as purchasing, legal, contract negotiations, etc.

These workshops require extensive preparation. It is recommended that the staff involved from both the service integrator and incoming service providers clearly understand the contractual commitments. Where there is interaction between service providers, it is common to work out the low-level operational interfaces in multi-party workshops, facilitated by the service integrator.

Cooperation is a strange concept to some organizations, but is becoming less so as more organizations adopt SIAM models or work as part of them. Most difficulty is found with incumbent service providers that are being inducted to a SIAM ecosystem and may seek to circumvent new ways of working and continue using established practices.

It is sometimes necessary to obtain support from a commercial manager or transition manager to help interpret obligations and ensure that all parties cooperate. This should be done with caution, as it can result in service providers rigidly following the contract and refusing to do anything unless it is fully detailed in it.

Define then onboard/onboard then define

Decide whether to define the contracts first, or the SIAM model first. Defining contracts first is a traditional approach with the definition of contracts proceeding onboarding when the lower-level interactions are worked out. The second option writes the contracts with a degree of discretion (high-level principles, but no detail), with the intention of defining and agreeing more robust parameters afterwards. Common practice is often a combination of both scenarios, depending on the circumstances.

Service integrator first/service provider first

Although the ideal has been stated as onboarding the service integrator first, it may be necessary to work with an incumbent service provider while building the SIAM model, with a new service integrator progressively taking over the management of the service providers.

This can cause difficulties, for example, with an incumbent's reluctance to change existing processes and tooling, or an attempt to subvert the incoming service integrator. If the 'service provider first' approach is necessary, it may need to be managed as a traditional outsourcing arrangement and be migrated to the SIAM model over time.

Postpone the development of governance

This option initially builds only project governance. The customer organization may try to have a major role itself in operating the SIAM model for some considerable period, with little perceptible governance in place.

Service providers often insist on retaining a direct management relationship with the customer organization rather than with the service integrator. This can help them to sell additional services or have a direct conversation when there are contract disputes.

Postponing the development of governance is likely to lead to a failure in implementing a SIAM model, with each service provider having different agreements with the customer and minimal coordination and collaboration between them. If this approach is taken, it is unlikely that there will ever be a fully effective integration layer, and the expected benefits from SIAM will not be attained. The governance framework for SIAM must be established during the Discovery & Strategy stage (see section **2.3 Establish a SIAM governance framework**).

3.3.2 Plan for appointing service providers

Selecting an external service integrator and agreeing a contract with it can take some time. On occasion, the customer might source the service integrator and the service providers simultaneously. Alternatively, the service providers might already be in place or be undergoing transition from legacy contracts before the service integrator role is confirmed.

While the service integrator is being selected, most customer organizations still have existing service providers to manage and existing services to be delivered. The customer organization needs to consider these existing arrangements alongside the future desired state, and develop a plan to ensure that business as usual (BAU) is not compromised as the desired state is developed. This is sometimes referred to as 'keeping the lights on'.

There is often a mismatch between existing support arrangements and contracts, and the desired future ways of working. Some contracts may be mid-term and not simple or cost effective to alter. There needs to be a pragmatic acceptance that minor or non-business critical service provider contracts will not necessarily reflect the new ways of working.

Non-compliant service providers

The service integrator can treat non-compliant service providers in a similar way to providers of commodity cloud services that are also unlikely to align their practices (see section **5.3.2.2 Dealing with incomplete or non-standard data**).

For example, it can attend the structural elements on their behalf, conduct any toolset integration for them, reformat their service reports into any new format, etc.

Planning needs to consider the customer organization's objectives, and also the likely reaction and action of other stakeholders. When planning to introduce a SIAM model where none exists, the effect on incumbent service providers, their willingness to change and their expected behaviors must be carefully considered.

Sun Tzu

As Sun Tzu identified in *The Art of War*,[28] there is no obligation on behalf of a strategic player to tell the whole truth.

"All Warfare is based on deception.

Hence, when able to attack, we must seem unable; when using our forces, we must seem inactive; when we are near, we must make the enemy believe we are far away; when far away, we must make him believe we are near ...

---

[28] *The Art of War* by Sun Tzu, Williams Collins, 2018.

> *Now the general who wins a battle makes many calculations in his temple ere the battle is fought. The general who loses a battle makes but few calculations beforehand. Thus, do many calculations lead to victory and few calculations to defeat: how much more no calculation at all! It is by attention to this point that I can foresee who is likely to win or lose."*

When considering incumbent service providers, the customer organization should address the Discovery & Strategy activities, assessing the service provider and assessing the service. An incumbent service provider's interests will be aligned to the probability of it having its contract extended for the equivalent or a higher contract value. This is especially the case where a service provider has an element of its own remuneration based on the value of new or extended business.

If a customer organization is requesting that a service provider exchanges a contract that commits them to delivering five years' profitable service, for a contract that is much shorter with less overall profit, the offer is unlikely to be well received. If on the other hand, the customer organization offers to exchange short-term for long-term commitment, they must be clear on the value to the service provider of what they are presenting.

All parties should consider their options. Considerations should include:

- Incumbent service providers:
  - o If the contract does not provide sufficient profit or adds complexity for the service provider, it may not wish to continue to engage with the customer organization.
  - o Contributing to a SIAM model will require insight into the services that go beyond typical contracts. It will involve additional obligations in the form of collaboration through structural elements and other interactions, as well as a requirement to contribute additional documentation and reports. It is important that these obligations are understood by each of the SIAM layers, and that each has considered how to approach their fulfilment.
  - o Whether there are strategic, commercial or market benefits from a SIAM engagement.
  - o Financial impacts.
- Service providers and the service integrator:
  - o Understanding how engagements within the SIAM model could impact on other services provided to their current and prospective customers.
  - o Assessing the commercial sensitivities surrounding competitive parties in the market. Where there are competitors that may be part of the SIAM model, there may be sensitivities or indeed policies on working together.
- Customer organization:
  - o Whether moving from a single service provider to multiple service providers (or alternatively, going from direct governance of multiple providers to using a service integrator) presents a threat, and how to mitigate this if required.
  - o Assessing whether retained capabilities are equipped to make the transition to a SIAM model.

An incumbent service provider has greater knowledge of the customer's services, of the customer's senior stakeholders and better access to them than an incoming service provider.

As a service provider, there is no obligation to be entirely open with the customer organization, any more than a service provider should expect a customer to report all contact with competitors. The management of messages and who is likely to do what in which circumstances can be a lively game that needs to be 'played well'.

SIAM without cooperation

A large organization operating on many sites had a set of established incumbents each supplying services within well-defined and largely separate areas. It decided to adopt a SIAM model. One of the service provider contracts was due for renewal, suggesting a suitable start point for migration to a multi-service provider model under a service integrator. A second large contract had some years to run.

It was decided to approach the second service provider, requesting a contract change. Part of the settlement was to give service integration responsibility to the second service provider. Although this service provider was an international tier one provider with immense corporate capability, it did not have a culture of collaboration or SIAM experience.

Theoretical 'Chinese walls' were constructed between the service integrator role and the second service provider's continuing services, to no visible effect. The service integrator failed to perform and was distrusted by all.

The success of a SIAM transition relies on development and maintenance of enduring, open and honest relationships with all parties – customer organization, service integrator and service providers. It is vital for all service providers to remain focused on providing the customer organization with end-to-end high-quality service, even when their contracts are mid-term and targets are not fully aligned to the customer organization's required outcomes.

Contribution to end-to-end performance

During the Plan & Build stage, a customer organization had a challenge to set up contracts for the onboarding of service providers responsible for one part of the end-to-end service. Although it was only providing one element of the service, it had to be committed to end-to-end performance.

The customer organization invited external consultants to help it define a service contract, including clauses that set out performance targets within several SLAs, all of which had commercial implications, and KPIs to measure performance of key indicators.

The contract with the service providers included incentives paid for consistently outperforming targets and service credits to be deducted for underachieving. The contracts also provided for targets to be 'stretched' to encourage improvements and innovation when they were consistently met for several months.

The service provider reported on KPIs and SLAs monthly. When a service provider underachieved on a KPI for three consecutive months, the KPI was upgraded to a defined SLA, which had commercial implications. This was one of the approaches the customer organization used to encourage service providers to contribute to end-to-end service performance.

Failures in providing a well-functioning SIAM model may happen because of a lack of consideration of the services required, or how they fit together. One size does not fit all! If the customer organization relies on the procurement team to simply bundle together vertically-integrated contracts for services, such as 'network' or 'desktop', the desired outcome will probably not be delivered. Commodity products such as hosting are often summarily outsourced to commodity providers. Services closer to the end user, critical business services or vital functions are more likely to be built and managed internally, as this is perceived to provide more control and introduce less risk for the customer organization.

3.3.3 Transition planning

As part of the bid that leads to the contract, an incoming service provider should:

- Propose an approach to manage its transition into the SIAM model
- Propose milestones by which the progress and success of the transition will be measured
- Identify and cost the assets to be introduced and any to be transferred
- Manage information on any staff to be transferred
- Develop and cost a resource plan for the transition and ongoing operation
- Identify risks and use those as a principal input to due diligence verification before signing the contract
- Identify and assign the resources required to deliver the transition, or at least where those resources can be obtained from and lead times for this
- Develop the outline plan for delivery

Appropriate bid material should be merged into the contract as part of a transition schedule. This makes the approach and milestones contractually binding.

If there are any services that are being transferred to an incoming service provider, they should form part of a transition plan. The level of detail will depend on the specific situation and the characteristics of the service that is being transferred. At a high level, the following elements should be part of a transition plan:

- An approach document describing a definition of the scope for each workstream and how the incoming service provider will deliver this.
- A schedule, using project management artefacts, that identifies the milestones, timing, task sequence, resourcing, product breakdown and interdependence. This is used as the basis for periodic progress reporting.

The customer organization and the service integrator should review the plan to make sure there are no overlaps or gaps, and to confirm that they are getting what they expect and that they know what is expected of them. Careful, rapid and timely review and feedback is essential to highlight assumptions and conflicting interpretations. These can be changed quickly and cost-effectively if identified at an early stage.

There is a material difference between the level of definition required for a bid and that required for delivery, but the two must be compatible. There may be different people involved in aspects of the bid and delivery. A degree of continuity is highly desirable, but more staff will be required in the delivery. The contract should allow for review and approval of the transition plan as one of the milestones.

The customer organization must prepare for and conduct a thorough review of the transition plans of the incoming service providers to ensure that:

- Subject matter experts (SMEs) from the retained capabilities are familiar with the transition plan
- SMEs from the service integrator are familiar with the transition plan
- Dependencies, assumptions and risks are all known and accepted
- The plan is compatible with the contract

The service integrator should develop an overarching plan that integrates all the separate transition plans from the new service providers, and any incumbent service providers.

Where there are interdependencies between services provided by different service providers, the transition plans from each of these service providers should be analyzed to ensure that they will work together and that there are no gaps or overlaps. It can be useful to share the plans between the affected service providers, and the contracts must allow this.

Often, a short period is contractually provided to allow for review. The customer and service integrator must be prepared and have reviewers ready and arrangements in place to conduct the review.

The review must be done with reference to the obligations in the contract with the service provider. This prevents a reviewer, with good intentions, assuming obligations that are not in the contract. Assumptions or a lack of clarity can distract from the main purpose, which is to deliver a good quality transition from the outgoing to the incoming service providers, on time, to cost and with continuity of service availability. The observation of a serious omission or issue should be addressed through contract change (which can potentially have implications on the funding and schedule).

3.3.3.1 Transitioning services

For any service provider already operating in a different SIAM ecosystem, the service integration requirements and obligations for any new ecosystem will need to be overlaid on or integrated with its current practices. This can add additional time and complexity.

All service providers, and the service integrator, need to consider the time and cost required for the discovery, knowledge transfer and design workshops for the incoming service provider and the parties with whom they must engage. It is common to manage the transition of each a service in a separate workstream, with a plan addressing:

- Mobilization and planning
- Documented data discovery
- Knowledge transfer (discussion, shadowing, observation in practice)
- Build
- Test
- Validation, operational and service acceptance
- Cutover
- Early life support
- Project close

The mode of transition can vary widely between services, and will reflect the bid and contracted approach. Possible modes include:

- Contract and service provider change: operations remain as-is
- Transform then transition: big bang
- Transform then transition: site by site/phased

The variety is vast.

Selected elements such as toolset integration, introduction of any new tooling and the adaption of processes consistent with the necessary interfaces are likely to be required from the first day of the service.

3.3.3.2 Resourcing transition

An experienced and capable transition manager with appropriate sponsorship and a supporting project management office (PMO) (see section **2.2.3.2 Project roles**) will be needed by each party: incoming and outgoing service providers, service integrator and customer organization retained capabilities. Everyone must have a thorough working knowledge of the relevant contracts.

Each party will need to have appropriate budget available. Sometimes the customer will pay some of the costs of the service provider, for example, to expedite the transfer or build integrations with a tool.

It is common practice for an exit schedule to provide a bare minimum of funded services and obligations, with additional support at a stated rate beyond this. This can lead to significant charges for exit support. Because at the time of writing the original contract, the service provider could not be sure of the nature of exit support that would be required to fit in with the successor's transition, exit schedules tend to be very flexible.

Outside a SIAM ecosystem, transition projects are often led by the incoming service provider, supported by the customer and the outgoing service provider.

The reasons for this are:

- There is no contractual relationship between the outgoing service provider and the incoming, so any relationships need to be mediated by the customer, who will have obligations to the incoming service provider to do so, and an exit schedule with the departing service provider to support this.
- The incoming service provider is responsible for managing the transition of its service(s), preparing itself and obtaining what it needs from others to achieve this.

For transitions within a SIAM ecosystem, the service integrator should be accountable for the transition of service(s). SIAM models have many service providers and commonly short duration contracts, which will lead to more frequent changes of service provider. It can make sense to retain a permanent transition capability within the integrator layer and customer retained capabilities.

The service integrator and incoming service providers need to define their relative roles and responsibilities carefully. Getting these wrong can negatively affect the success of the new service and create disputes.

The advocated approach is:

- The service integrator is responsible for representing and assuring compliance to defined governance so that the required end-to-end service integrity is achieved
- Service providers are responsible for delivering services to the customer organization
- The service integrator is responsible for delivering the customer's transition obligations, including ensuring any outgoing party delivers what is required of them

Throughout the transition project, an established project management method should be used by the transition team. This should include supervision by a governance body receiving regular reports and subjecting the transition managers to appropriate scrutiny, giving them guidance and direction where appropriate (see section **2.2 Establish the SIAM transition project**).

Transition managers

The skills of transition managers can be both deep and broad. This means that the good ones capable of managing a large implementation are likely to be expensive. The bad ones are considerably more expensive in the end, as they can lead to missed objectives, rework of plans and missed dates resulting in unplanned and expensive continuation of current services.

The organization of the transition team will differ between the service integrator and the outgoing and incoming service providers, reflecting their roles. Incoming service providers are likely to have some functional workstreams (for example, externally sourced HR) and some service-based workstreams (for example, networks or data center). Any task that is broken up for the convenience of management will need to have someone responsible for bringing the components together coherently. This is commonly the role of the organization's transition manager.

Each party should appoint a transition manager to oversee its organizational concerns and manage coordination with other parties.

Outgoing service providers normally call this role an 'exit manager'. All transition managers should meet regularly throughout the transition. The service integrator should chair the meetings and set the agenda for what happens in the transition, in line with the incoming service provider's transition plans.

3.3.3.3 Handovers

In the Plan & Build stage, it is important to make plans for the handovers that are inevitable between incoming and outgoing service providers once a SIAM model is established. Handovers are a necessary aspect of a multi-service provider support environment. Often, problems are encountered when tasks are handed over from one person to another, or from one team to the next. When a handover is done improperly, friction frequently results and the efficiency and effectiveness of everyone involved will be affected.

A problem often encountered in complex environments is improper handover of operational support activities or responsibilities. When the SIAM model has been created hurriedly, key personnel have left quickly or the service environment is immature, there may not have been an appropriate transition period allowing for designated people to become familiar with their responsibilities. To prevent such challenges, it is good practice to ensure that as much time as possible is factored into the transition activity in the Implement stage.

These measures should be considered:

- List all the handover activities and responsibilities required
- Prioritize activities from the most important to the least important
- Define the positions and names of all individuals who are part of the interaction or are in the chain of command
- Define the coordination required between the person handing over the responsibilities and the person who will be the recipient
- Where support is being passed between layers or providers, it is important to produce an appropriate level of documentation in the form of processes, procedures and work instructions
- The incumbent party should create handover reports and provide these to the newly responsible individual or team, which includes:
  - Summary of tasks, duties and responsibilities
  - Identity of the recipient of the handover, or who will be taking over the tasks, duties and responsibilities
  - Timeline of handover, from the time to commence and the anticipated handover period
  - Activities to be completed during the handover
  - Any notes or special points of interest that might support knowledge

---

Goodwill

There will always be things that emerge after the incumbent service provider has gone, regardless of how much planning has gone into it.

In most cases, the previous incumbent will provide advice (but not actual support) on a goodwill basis, but it is unwise to assume this will happen, as it may not always be the case.

Created in the Plan & Build stage, documentation of handovers becomes important during the Implement stage. By clearly defining the roles and responsibilities of all elements of the SIAM model, handovers will be easier to conduct and it should limit issues within the ecosystem implementation.

3.3.3.4 Outgoing resources

Despite the best planning, there can be considerable issues during the transition of a service to the new SIAM model from incumbent providers. As has been previously mentioned in the Discovery & Strategy section, without careful planning and a coherent strategic approach, transitions can encounter problems if assumptions are made around how outgoing disaffected service providers and employees might behave (see section **2.6 Define the strategy**).

It is unusual for service provider exit plans to have detailed clauses, as at the time they are created the precise nature of the exit will not be known. Consequently, outgoing service providers may be unwilling to engage and cooperate with the incoming service providers and the service integrator, as their focus may be on maintaining business as usual (BAU), and not supporting the transition to the SIAM model.

Depending upon the specific circumstances, employees in the UK and EU could be covered by the Transfer of Undertakings (Protection of Employment) Regulations 2006 (SI 2006/246), commonly referred to as TUPE, or the Transfers of Undertakings European Union Directive.[29] Specific, local employment legal guidance should be sought at the earliest opportunity (see **Appendix D: Staff displacement legislation**).

TUPE[30]

The Transfer of Undertakings (Protection of Employment) Regulations - 2006 (SI 2006/246), known colloquially as TUPE, are the United Kingdom's implementation of the European Union Business Transfers Directive. It is an important part of UK labor law, protecting employees whose responsibilities are being transferred to another business.

The regulation's main aims are to ensure that, in connection with the transfer, employment is protected (for example, substantially continued) and:

- Employees are not dismissed
- Employees' most important terms and conditions of contracts are not worsened
- Affected employees are informed and consulted through representatives

---

Employees and roles that come under the scope of any legislation should be identified early in the transition lifecycle to avoid unpleasant surprises. There is always a risk that affected employees do not declare their intentions until the last possible moment, late in the transition lifecycle, which can be disruptive to the transition plan.

3.3.3.5 Outgoing service providers

As part of the SIAM project, the customer organization must choose whether to keep or phase out existing service providers. It is necessary to ensure that there are adequate processes in place to coordinate the decommissioning of proprietary contracts, while managing service providers that will remain.

Poor incumbent service provider management can lead to significant risks. Strained relationships often lead to compromised support and relationships that invariably slow the transition to the new SIAM model.

Outgoing service providers may not share the same agenda as the customer organization, which will be focused on a smooth handover to the new SIAM model, and may not 'play fair' or participate in any activities beyond its contractual commitments. This can take the form of lack of engagement as outgoing service providers wish to protect:

- Revenues
- Competitive edge
- Intellectual property (IP) rights
- Employees

---

[29] *www.legislation.gov.uk/uksi/2006/246/contents/made*.

[30] For more information visit: *www.gov.uk/transfers-takeovers*.

This needs to be recognized by the service integrator, and mitigations provided for in the detailed transition plans and agreed by all parties. Incumbent service provider management can be performed as part of supplier management within the service integration layer, but needs to integrate with contract management in the customer organization's retained capabilities.

Mechanisms for escalating issues need to be in place. The only leverage the customer has on an outgoing service provider are existing contract provisions, the threat of loss of potential future business and reputational risks. The last two can be very powerful. Good practices in incumbent service provider management include:

- Secure and maintain an incumbent service provider's goodwill during the transition period.
- Maintain relationships and service provider willingness to participate where relevant in the tendering of new, additional or updated service components under the SIAM model.
- Develop an interim operating model setting out how to interact with service providers during the transition stage until SIAM practices have been established. This is sometimes referred to as an interim service plan and is particularly useful to manage an incumbent service provider that is going to be replaced as part of the final SIAM environment but may need to change some ways of working during the transition.
- Provide a single point of contact and ownership for the support of all incumbent service providers. The defined governance model should provide guidance regarding the ongoing approach for engaging and managing them throughout the transitional activities (see section **2.3 Establish a SIAM governance framework**).

Develop a negotiation strategy to help mitigate the risks and associated costs of managing any outgoing legacy contracts. This strategy can include:

- Identification of new opportunities for the incumbent service provider to reduce loss of revenue.
- Identification of threats and costs to be resolved before contract exit.
- Create draft contract change notices for existing agreements with incumbent service providers to clarify and encapsulate their obligations and ensure the safe transition or disposal of agreed assets and information.

Exit costs during the migration of services to new service providers can be mitigated by negotiating clear and firm charges for termination assistance and exit charges for stranded costs and assets (see section **3.1.3.10 End of contract**). Retaining positive relationships with incumbent service providers will help to maintain service quality during the transition and preserve the relationships for possible future engagements.

3.3.3.6 Planning for service retirement

The retirement of a service is a major change and should be managed accordingly. Management of a retiring service involves:

- Ramping down and transferring work in progress. This focuses on open incident, problem, change and service request records. Ramp down commonly includes progressively shutting down the ability to request services in advance of transition and implementing a change freeze.
- Asset cleansing and transfer or decommissioning.
- Data extraction and transfer (for example, configuration and asset registers).

- Migration or archiving (recoverable, accessible as required) business data.
- Contact center communications redirection and the transfer of associated communications ownership.
- Implementing engineering changes to permit the control of the service by the incoming provider.
- Paying service providers for the project services required (where not otherwise funded).
- Reallocation of service personnel.
- Termination of service provider contracts no longer needed.

3.3.3.7 Accommodating inflight projects

It is likely that there will be some projects that are in various states of deployment during the SIAM transition, which may have variable impact on the incoming service providers and the service integrator. These projects may be triggered by, be the trigger for or be independent of the SIAM transition project. Examples include:

- Moving or rationalizing data centers
- Reorganizing the service desk
- Business and IT service continuity projects
- Application rationalization projects
- Service management tool upgrade or replacement

The incoming service providers and the service integrator will need to have visibility of these projects to be able to assess their impact on the planned SIAM transition (possibly including test results, rollout schedules or project issues). All inflight projects should be identified during the Discovery & Strategy stage, and further assessed during the Plan & Build stage, including their status, who is responsible for them before and after implementation, and any associated assumptions and risks (see also section **2.5.7.2 Inflight projects**).

Unless specified in the contracts with the outgoing service providers, there is no guarantee that they will take any responsibility for continuing these projects. If possible, larger projects should be completed before a new service provider joins the SIAM model. Alternatively, it might be possible to plan a transition of responsibilities for inflight projects as part of the SIAM transition, aligned to key milestones and depending on the project stage, even if this means that the 'go-live' date does not directly align to the SIAM transition. For projects in their early stages, it is often better to transition to the new service provider, whereas for projects in their later stages, it may be better to allow the incumbent to conclude the project.

One option is to stop all non-urgent inflight projects at the end of the Discovery & Strategy stage. Although this will affect the timing and delivery of outcomes of these projects, it can greatly simplify the Plan & Build, and Implement stages of the SIAM transition project, and significantly reduce the risks to all these projects.

3.3.3.8 Operational acceptance criteria

When transitioning to a SIAM model there will be a series of transformational activities:

- Fully document the SIAM model to be delivered and communicate to all stakeholders, ensuring that the model and any underpinning principles are clear to all.
- Prepare and communicate any new or amended contracts.

- Prepare and communicate any service provider transition or exit plans.
- Form the necessary teams (centrally as well as locally in case of a distributed organization) to help support commitment, mobilization and momentum throughout the SIAM transition (see section **2.2.3.2 Project roles**).

With any complex transition, delays and issues are inevitable. It can be helpful to use an interim operating model to cover the period during which service providers are transitioning in and out. If this is the case, all stakeholders need to have their expectations set regarding how the interim model will work and what it might mean for them.

Clear and consistent instructions are needed to make a link between legacy contracts and the new SIAM model, including a list of service providers, contact details and information about contractual obligations. The service integrator, once appointed, needs to provide additional guidance on accountabilities and responsibilities within any interim model.

Where possible, include some flexibility in contract start and end dates. Maintaining relationships with outgoing service providers is important to support this flexibility. Increased costs from short-term contract extensions or parallel running may not be avoidable, but should be minimized where possible.

Interim operating model

A global enterprise in the automation and robotics industry transformed its legacy second-generation outsourcing contract for infrastructure services into a SIAM model. During the transition, the following situation occurred.

The service desk was already consolidated and was the single point of contact for user issues. There were various service groupings behind the service desk, with different service providers carrying out day-to-day activities. Supporting each service provider was a function monitoring service performance and stepping in as required. A service integrator was appointed and stepped in to deal with cross-service provider issues. A global service management software provider was used to provide the SIAM toolset.

There was a large variance between the operating models of different countries. This extended to toolsets, processes and service management practices. All country organizations had their independent infrastructure operating models, which did not match the new, global, SIAM operating model. Although the global service desk was already available, the infrastructure services were still being managed by the individual countries, with some transitioned into the SIAM model and others waiting to be transitioned.

In a global transformation, this is a typical situation. It is often the case that different parts of the organization and their service providers will not transfer their technology services from current to future mode at a single point in time.

An interim operating model was created to allow service provision to continue during the transfer of services into the new model:

- The global service desk became the new single point of contact for all user interaction regarding infrastructure
- A new global service portal provided the single point of access for service requests and incident logging

---

- The service desk triaged incoming issues and passed those issues that could not be resolved to the appropriate service provider
- All incumbent service providers were moved to the new global service management tool
- For issues that required joint resolution by the incumbent service provider and the SIAM team, an escalation path was defined and actively managed

The purpose of an interim operating model is to allow day-to-day service provision to continue, minimizing risks during the SIAM transition.

The criteria for formal acceptance into operational support needs to be agreed and included in the contracts. This can take the form of an independent review of the available artefacts for each process area or service discipline, linked to the contractual obligations and critical success factors (CSF)s. For example, for service continuity:

- Are there well-defined and tested service continuity plans for each service?
- Are the business-critical services identified?
- Are the service continuity invocation processes understood for each service?
- Are the service continuity escalation points documented, available and tested?
- Are the links to crisis management understood?
- Is there an overarching service continuity plan that incorporates those from the service providers?

Checklists covering all aspects should be detailed and specific. These should be linked to the risk management approach used for the transition to the SIAM model, as failure to achieve any of the criteria will result in a risk to the specific and the end-to-end service.

The service integrator can track progress against the checklists throughout the Plan & Build, and Implement stages, and initiate actions when progress is causing concern. Tracking progress throughout helps to avoid surprises and supports any necessary re-planning.

Sometimes, operational acceptance can take the simplified form of a formal sign-off by an individual in each service provider, as well as the service integrator. However, this can turn into a 'staring contest', waiting to see who will blink first and raise a specific concern on an element of service.

Although it is crucial to ensure that each service provider is ready to operate its services, the service integrator has overall responsibility for operational acceptance. It must consider acceptance of the ecosystem as a whole, including the readiness of all service providers, itself and the customer's retained capabilities.

3.3.3.9 Managing the risks of transition

The Plan & Build stage plays an important role in preparing for a successful Implement stage. Continuity of existing service provision is a critical requirement for the transition of services to the new SIAM model. The uncomfortable reality is that this cannot necessarily be guaranteed (see also section **3.3.3.10 Phases of service levels in transition**).

Before and during the transition, those responsible will seek to identify an approach to accomplish the transition that minimizes the risk to service continuity within the constraints of cost and time. Extending the duration of the transition may not reduce the risk, as elements can be difficult to sustain once changes have been started. This is particularly so where declining staff morale, or staff leaving, is an issue.

The incoming service providers will have been told about the service to a level of detail during the bid phase, sufficient to prepare their bid. The degree of definition required to build and operate the service is significantly greater. The gap is addressed through a due diligence phase, incorporating documentary research followed by conversation. It is common to use a series of workshops to facilitate this, supported by work shadowing and observations of operations in practice.

There will be a degree of residual risk that the incoming service providers have either missed elements or misinterpreted what they have been told. This risk is compounded where:

- The duration of knowledge capture is short and overall schedules are infeasible
- Cooperation of outgoing service providers is low
- Staff leave before, during or after transition
- Intellectual property (IP) and commercial rivalry between the service providers impedes progress
- Transition is poorly planned or executed
- Communications are slow, and staff are unresponsive
- Resources are not made available
- Sponsorship is poor, indecisive and unsupportive
- Relationships are poorly managed

3.3.3.10 Phases of service levels in transition

The levels of service performance that may be expected around a transition are illustrated in **figure 20**. The severity of impact varies widely with the maturity of the service, age of infrastructure or the circumstances and the competence of the staff involved.

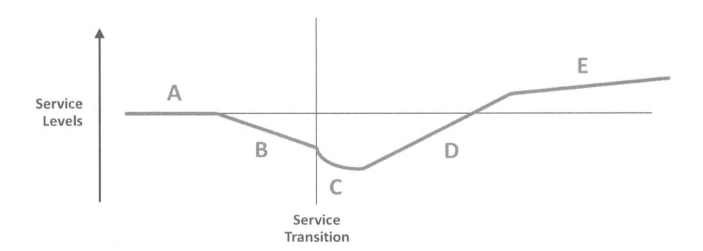

Figure 20: Service levels in transition

The phases are:

A. Outgoing service provider's baseline service level
B. In the approach to transition, the outgoing service providers withdraw resources and service levels decline
C. Post-transition disturbance as the incoming service providers assume service with immature operations methods and staff, and disjointed tooling
D. Service stabilizes, and performance issues are resolved
E. Stable operations are achieved with continual service improvement

3.3.4   Commercial management

Each of the parties to the transition (including incoming and outgoing service providers and the customer organization) will focus on their own objectives. These are likely to include:

For the incoming service provider:

- Receipt of milestone payments on time
- Management of assignment profitability
- Customer satisfaction as a prelude to a successful service and relationship

For the customer organization:

- Acceptable levels of service continuity
- Contented (or minimal impact to) users
- Transition achieved on time, within budget and to business case
- Obligations delivered, a good service established

For the outgoing service provider:

- All debts settled
- Costs contained and profitability maximized
- Key staff retained
- Reputation maintained, perhaps with the hope of more work from this customer in the future

The parties will also be concerned to a greater or lesser degree with the satisfaction of their obligations and the delivery of dependencies by others. The outgoing service providers will be interested in this only to the degree that it contributes to the achievement of their objectives. In practice, a customer has very little commercial leverage over an outgoing service provider. Relationships, appeals to personal pride and professionalism, and references can be much stronger influences if there is little immediate prospect of more work.

The orderly management of compliance to contract can be usefully supported using a supplier and contract management information system (SCMIS, see section **5.3.4 Ongoing service provider management**). This can be used to store all relevant documents, trackers and reports related to all service providers. These take a contract-centric view and should produce a high-level view of reminders to assure that obligations are delivered. Such systems rely on good contract management processes and appropriate retained capabilities within the customer organization.

The management of external contract transfer can be particularly troublesome. The ownership of assets must be established, together with software licenses and maintenance agreements. Just finding the agreements can be a major undertaking, particularly if they are not recent. Permission must be obtained to share confidential agreements. Contracts must be novated or reformed, and management transferred. Operational contacts must be established, and processes and tooling for data exchange implemented. This activity is necessary to ensure that the incoming service provider has assurance that all components of the service are under effective maintenance agreements and that the ownership of assets is beyond dispute.

3.4 Integration recommendations

One of the key objectives of service integration is to provide a 'single source of truth' that can be achieved only through data availability, consistency and transparency. If there are multiple toolsets with different data and reporting criteria in use, it will be difficult to manage the overall ecosystem effectively. A tooling strategy will outline how to deal with this (see section **3.1.9 Tooling strategy**).

Creation and maintenance of an integrated configuration management system (CMS) and associated configuration management databases (CMDBs) can be particularly complex, especially if a single toolset has not been chosen.

The service integrator must decide what data it needs to capture to carry out its responsibilities, and should capture and maintain the data in the relevant toolset. Toolset integration mechanisms may need to be developed to capture initial data from the service providers and update it when changes are made. This often requires data to be translated and/or transformed. For example, time and date formats may be different between different service providers. A common data dictionary should outline standards for data, including impact, priority, product and operational definitions to be used across the toolsets.

The customer organization needs to ensure that the service provider contracts contain appropriate requirements for data integration and provision, to support consolidated reporting by the service integrator. If the service integrator is appointed before the service providers, then they can help to define these requirements.

The service integrator will focus on service provider performance by measuring it against individually defined targets. Although the individual service providers will need to understand and demonstrate their own performance, the service integrator needs to collate data to allow an integrated, end-to-end view of service performance. This can be drilled down to a more detailed level as required. Data drill down is a key functional requirement for a service integrator's 'service dashboard'.

As part of adopting SIAM, multiple reports and different reporting formats may need to be streamlined and aggregated to form a common reporting model. The customer organization and the service integrator should have real-time visibility of service performance for both end-to-end services and, where necessary, supporting sub-services. There should also be a drill down capability to track, for example, incidents or service providers that missed service level targets.

Processes and tools in a SIAM ecosystem must provide end-to-end quality and performance measurements across the service providers. This end-to-end measurement and service assurance capability is critical for streamlined and effective delivery of value to the business. Measurement of end-to-end performance can be challenging, particularly if the service levels vary by service provider. It requires an understanding of the end service received, the ability to measure the availability and performance of that service, as well as of the underpinning services and service elements.

Reports are provided to the customer organization through the service integrator only. This provides ownership and accountability for the data, and allows the service integrator to take a proactive role.

One of the more complex aspects of service integration is the financial management of multiple service provider contracts, including regular activities such as the validation of invoices and the justification of allocated charges back to a specific organization. This can be time consuming and result in disputes. To facilitate this, an integrated enterprise view of invoices and resource units across all service providers should be available in the 'service dashboard', but only to a restricted audience.

3.5 Applicable SIAM practices

> **Practice definition**
>
> *"The actual application or use of an idea, belief or method, as opposed to theories relating to it."*[31]

Within the SIAM Foundation BoK, four types of practice are described:

1. People practices
2. Process practices
3. Measurement practices
4. Technology practices

These practice areas address governance, management, integration, assurance and coordination across the layers, and need to be considered when designing, transitioning or operating a SIAM model. This section looks at each of these practice areas and provides specific, practical considerations within the Plan & Build stage. Note that the People and Process practices will be combined and referred to as 'Capability'.

3.5.1 *Capability (people/process) considerations*

The Plan & Build stage builds on outputs from the Discovery & Strategy stage to complete the design for the SIAM model and create the plans for the transition. During this stage, all plans and approvals are put in place before the Implement stage begins. This involves adding detail to critical SIAM artefacts. To support this, it is important to ensure that the requisite capabilities exist within each of the SIAM layers.

Capabilities are required in these areas:

- Full business case
- SIAM model
- Selected SIAM structure
- Structural elements
- Governance model
- Roles and responsibilities

---

[31] *www.lexico.com/definition/practice.*

- Appoint service integrator
- Appoint service providers
- Collaboration model

3.5.1.1 Practices for integrating process models

Section **3.1.4 Process models** discussed the importance of process design within a SIAM model. Practices for integrating processes across service providers include use of an enterprise process framework (EPF) to provide a mechanism for harmonizing process disparity and managing the associated complexities inherent within a SIAM model. The framework chosen should define architected views to support definition of principles for alignment, process interdependence, governance, organizational operations and process traceability. This will enable clarity and visibility.

The EPF provides a method for abstracting the organizational processes and their various elements. **Figure 21** below shows an example of the key elements of an EPF framework. It includes:

- Process definitions (the outermost elements)
- Enterprise alignment and governance (in the center)
- Process intersections and dependencies (the arrows)
- An underlying process architecture that synchronizes the elements

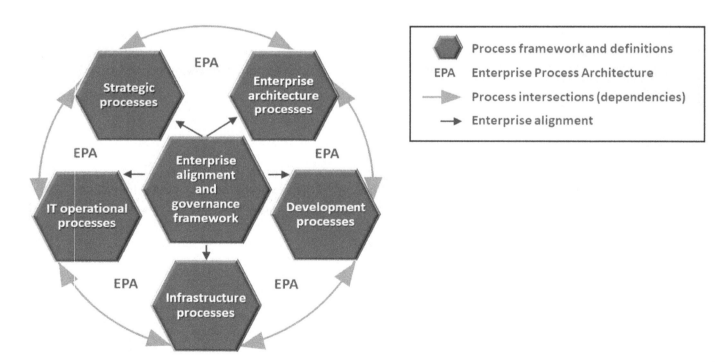

Figure 21: Defining an EPF

The focus should be on the process outcomes and designing an approach to support continual improvement.

Process models must be established. The service integrator must ensure that processes are coordinated across service providers, are traceable to policies and do not contradict each other.

Process outcomes versus approach

SIAM does not mandate the approach of the individual service providers in the model. Instead, the service integrator is responsible for joining the links in the chain by providing process policy, high-level process flows, process inputs and ensuring that these will deliver the required process outcomes. Each service provider is responsible for turning those inputs into outputs within an agreed timescale and to an agreed standard as indicated in the process policy.

For example, for the change management process, the service integrator should not mandate the use of ITIL or COBIT (or any other approach) to the service providers. Doing so would restrict the service providers and compromise the economic specialisms and unique offerings that were the basis for selecting them.

However, there does need to be a policy and process flow showing in which circumstances and how, the service provider needs to engage with the service integrator and/or other service providers or the customer organization. The end-to-end high-level processes, including the interactions between the parties, should be defined in the contract and compliance written into the obligations, but low-level procedures within a service provider's sphere of operations should be left to them to define.

3.5.1.2 Process establishment across the ecosystem

The enterprise process framework (EPF) defines objectives, governance and alignment strategies as actionable process elements. This enables organizations to extend the enterprise process elements to their specific needs, and integrate them within their own processes.

Change management process

A customer organization defines a change management process as an enterprise standard to control technology changes in a consistent manner. This would include a set of standard roles, tasks, work products and a common workflow.

The software development service provider would use the enterprise change management process to meet the requirements for change management of software.

The infrastructure service provider would extend the same enterprise change management process to fit within its ITIL-aligned definition of change management, focused on control of change to production.

An EPF introduces an actionable process at the enterprise level and leaves specific implementation to the individual organizations. Each of these organizations can apply an actionable process that integrates across the enterprise in a consistent and seamless manner.

The benefit of this approach is that it can be used for both greenfield implementations and ones where there are incumbent service providers. It also allows for flexibility within the service providers, and avoids rigidity and a 'one-size-fits-all' approach. The EPF can be used to understand how and to which standards service providers need to link, and how to apply that within their own support environment.

3.5.1.3 Abstraction of standard process elements

Abstraction is concerned with reducing complexity by filtering out the fine details. An EPF abstracts the common and critical process elements, and provides visibility into the common elements, ensuring they are not lost in the details of day-to-day tasks or subsumed by a single provider. With abstraction, the common elements can be extended by individual organizations to integrate governance and enterprise standards into their processes without encumbering their own organizational needs.

Enterprise architecture governance

Consider the customer organization's EA standards around technology adoption, which should be defined in governance and standards.

A software development provider can expand the EA governance criteria to meet internal software development standards around a specific technology. Meanwhile, an infrastructure group would expand the same EA governance criteria along the lines of production hardware and operating systems.

For each individual service provider, the merging of its standards with the EA governance is transparent, yet its commonality can be traced to the originating source in EA. This enables each organization to build upon its own needs using the EA governance, while governance is seeded throughout the organization for managing changes to the enterprise technology landscape.

3.5.2 Measurement practices

Within the Plan & Build stage, a performance measurement framework should be defined. This will provide a consistent approach for systematically collecting, analyzing, using and reporting on the performance of the service providers and the service integrator. This framework is a management tool for the service integrator that will enhance the management and reporting of service providers by measuring the achievement of expected outcomes. Ultimately, measurement supports informed decision making.

Trust is a precious commodity in a SIAM ecosystem. It is often better to empower service providers to use their internal tools, processes and procedures (which have been optimized for their organization). Typical enterprise service management (ESM) metrics and measurements do not necessarily apply in a SIAM ecosystem. Understanding incident volumes, although important, may not be useful for understanding service, process and service provider performance. With the rise of big data and business intelligence, businesses are constantly asking for more insight, more business context and improved decision-making capabilities.

A 'trust and track' method of measurement is required so that the service integrator does not get drawn into operational reporting by measuring activities outside its remit to govern or manage. With its focus on governance, collaboration and business value, the service integrator should put in place targeted metrics and measurements that aim to:

- Assess a service provider's ability to do what it says it will do. There should be checks and balances in place to ensure that service providers are operating appropriately within the SIAM model, according to policies, standards and controls, even when they use their own service management tools.

- Measure the integration of processes, including identifying bottlenecks and service providers that are not adhering to the vision for the SIAM model.
- Understand how a service provider is participating within all levels of the governance framework, in particular process forum and working group attendance and behavioral issues.
- Monitor key contractual metrics such as service levels, to apply corrective and preventative actions.
- Quantify the quality of the relationships between the service integrator and the service providers, as well as between service providers themselves (see section **3.1.8 Collaboration model**).
- Identify improvement opportunities across services, processes, frameworks and tools.
- Assess the efficacy of the SIAM governance model, including handling risks, issues and escalations at the lowest possible levels (thereby avoiding pressing the 'panic' button and empowering lower levels to make decisions and remove roadblocks).

3.5.2.1 Governance of measurement

From the SIAM model and the governance structure control points that have been identified, detailed information needs to be provided on the following:

- **Metrics ownership**: define who owns the metric definitions as well as who owns the responsibility to measure and report on the metric. This also needs to specify the recipients or consumers of each metric.
- **Measurement**: specify how each metric can be measured as well as its frequency. Keep in mind that measurement and reporting frequencies could be different over services and providers, and will need to change over time as the model evolves.
- **Data**: define a data retention policy, including where details of metrics will be stored and for how long. This might be common for all data points or it might be different for each metric.
- **Automation**: any metric collection and reporting that is automated needs to be described in detail with technical architecture, measurement and calculation logic, polling intervals, customization capabilities and any dynamic querying/reporting capabilities.
- **Design and definition**: define clear guidelines on the objectives and business value expected from the metrics. There should also be a review and update cycle defined for each metric.

3.5.2.2 Develop a basic results chain

To demonstrate how activities and outputs are expected to lead to the achievement of results, it is good practice to develop a basic results chain. This results chain helps to define how resources (at all layers), activities and outputs relate to the customer organization's strategic outcomes.

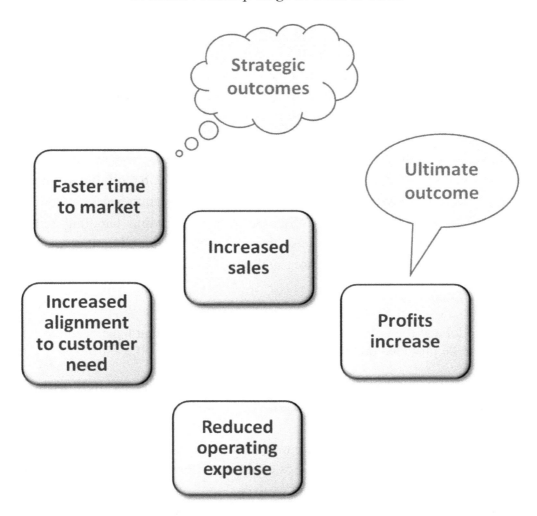

Figure 22: Results chain

The results chain, as illustrated in **figure 22**, is a qualitative diagram of the relationship between initiatives (inputs) and outcomes, and the contributions (connections) between the two. Following on from this, a benefits realization plan can be produced, which is a time-based, quantitative schedule of the benefits (see section **2.2.5.2 Benefits realization management**).

To do this, define the measurements for each initiative and desired outcomes, and iterate between the two. An outcome is a measurable, valuable result to the business that advances its purpose and goals. To be achievable, an outcome must be valid (for example, measurable), have all necessary conditions in place and have sufficient contribution. Think of maturity, which is something that grows over time. It is not a binary (have it or do not have it) quantity.

Results chain

Warning: A thoroughly worked-out results chain can be big!

An extensive program may include multiple initiatives and outcomes. If this is shown to a senior manager with a short attention span, it could prompt a negative reaction. Some people are averse to complexity and like to see simple, clear thinking. They also, however, like to see that someone has taken care of the detail.

A results chain should not be confused with a communication aid. It is a working document that helps to get the details straight. In general, tell the story in simplified words and pictures abstracted from the results chain, which can be appended to any final presentation, if it is included at all. It is hard to tell the story completely and accurately without having done the detailed work first.

The results chain is read from left to right and right to left. Left to right drives the thought process, 'if I do this, I get that result'. Right to left helps to ensure that the necessary conditions are in place, 'If I want to get this … I must have A, B, C in place in sufficient quantity'. The results chain can take on several levels of granularity. The picture can be filtered to show the appropriate level. All elements in the view should be represented at an equivalent level.

To create the results chain, the outcomes and initiatives should be outlined at a basic level. The contributions should then be depicted and can be moved around. From here, some editing should take place to tidy up, such as eliminating duplicate paths and contribution lines. Finally, number the elements so that they can be referred to uniquely. Some software tools are available that will help with numbering the elements.

It is best to draft the results chain first and then polish it for presentation. By the time the diagram is complete, all the necessary conditions are in place and the contributory links should be clear and credible.

Table 8: Elements of a Results Chain

Ultimate outcome	**WHY?** Why do we do this activity or provide this service?
Strategic outcomes	**WHAT?** What do we expect to see or hear as a result of our outputs and activities? **WHO and WHERE?** Who do we need to engage?
Intermediate outcomes	**HOW?** • The outputs, activities and inputs are effectively the operational elements required to achieve the strategic outcomes. • Which intermediate outcomes do we need to achieve the expected strategic outcomes?

Initiatives	Which key activities …
	Do we need to undertake to effectively contribute to the strategic outcomes?
	These are commonly delivered as projects or other major elements, such as a software system (hardware, software, services, operations, people) that can be costed.

An approach to developing a results chain for a SIAM model is summarized in **Table 8** and described in more detail below. The process is shown as a sequence, with a series of iterations as the objectives are understood through discussions with stakeholders that will be delivering the elements.

3.5.2.3 Principles of measurement

Principles should be established to allow the service integrator to measure performance on a consistent basis across the ecosystem. The following five principles should be considered:

- Outcomes and results must be clearly defined:
  - o Define the relationship between related measurements, and where there are competing factors, use hierarchies or categories to articulate criticality.
- The performance measurement system, including data collection, should be simple and cost optimized:
- o Wherever possible, data should be automatically provided to the service integrator, using a 'push' or 'pull' mechanism, depending on the capability of the toolsets.
- The performance measurement system should be positive, not punitive:
- o Despite the collaborative end-to-end focus on outcomes, measurements should not be incapable of revealing individual failure or success to allow follow-up action to be taken. It is an imperative in quantifying customer value that individual service provider measurements are available.
- Performance indicators should be simple, valid, reliable and relevant to the activity or process being measured:
- o The indicators must be understandable and drive desired behavior. For example, an indicator titled 'quality' is open to interpretation, whereas 'percentage of changes submitted that met the quality standard' has a clear meaning.
- o Typically, a higher value denotes better performance.
- Performance indicators should be reviewed and improved on an ongoing basis:
- o It is only by gaining experience measuring performance that you can refine and improve it.

During this stage, it is important to document measurement design and calculation to ensure clarity and transparency between all parties. It is a good idea to include a statement of the intent of the measurement, including the outcome that it is designed to drive, mapped to the strategic outcomes from the SIAM model. This helps to provide a foundation for future improvement. The service integrator should create and maintain a register containing all measurements and their status.

3.5.2.4 Choosing the right measurements

The right measurements are based on what *should* be measured (actual business outcomes) rather than what *can* be measured (incident resolution times, etc.). What is measured should evolve over time, as knowledge is gained that highlights the need for more and different data and information. The measurements should include, as a minimum, the performance targets that the service providers must achieve, as defined in their contracts or targets.

A balance should be achieved between outcome-focused measurements and behavioral measurements. Measuring outcomes alone can drive unintended behaviors. The design of drill-down reporting is also important to allow a balance between detail and focus on end-to-end outcomes. Use a balance of 'leading indicators' and 'lagging indicators' as each addresses measurement of an outcome from a different perspective (see also sections **3.1.3.6 Service credits** and **3.1.3.7 Incentives**).

Quantitative indicators are numeric or statistical measures that are often expressed in terms of unit of analysis (the number of, the frequency of, the percentage of, the ratio of, the variance with, etc.).

Quantitative indicator

A common quantitative indicator is whether an event was on budget, on time and within a service level, with an explanation of any significant variances.

Qualitative indicators are judgment or perception measures.

Qualitative indicator

An example could be the level of satisfaction reported by service users (how they 'feel' about the service).

Qualitative indicators may be quantified. For example, we may quantify the number or percentage of people who felt the service was excellent, good or bad. It is a common myth that quantitative indicators are inherently more objective than qualitative indicators. What is important is that we have a balanced set of quantitative and qualitative indicators to measure progress or performance.

Common practice suggests that it is reasonable to select three indicators for each performance measure (one quantitative, one qualitative and one discretionary that is often concerned with process/policy adherence).

Although there is no clear format or magic formula when selecting performance indicators, the following criteria can be used to determine the most appropriate indicators to measure performance:

- **Validity:** does the indicator allow you to be precise in measuring the results (quantity, quality, timeframe)?
- **Relevance:** is it relevant to the activity, product or process being measured?
- **Reliability:** is it a consistent measure over time? This is particularly important when selecting quantitative indicators.
- **Simplicity:** is the information available and will it be feasible to collect and analyze it?

- **Cost optimal:** does the benefit of collecting and analyzing the information exceed or neutralize the cost?

Another commonly used technique is to ensure that all measures are SMART:

- **S**pecific: well defined and clear
- **M**easurable: can be measured without overhead
- **A**greed: the parties have agreed on these measures (and who supplies them, and how)
- **R**elevant: support the outcomes
- **T**ime bound: the measurement is over a defined period of time

---

Measurements

An organization that recently implemented a SIAM model saw significant benefit from defining, documenting and automating the measurement and presentation of key end-to-end service levels. As a government organization with significant media presence, it was important for its organizational change management (OCM) to be transparent and oriented to provide a positive customer experience. This driver set the tone for how reporting and measurement was performed.

Definition of the key end-to-end service levels gave all service providers clarity of purpose, and set the focus on the customer perspective of service delivery. Documenting and being transparent about how these service levels were calculated promoted collaboration between service providers and the service integrator. Additionally, it provided a platform for amending the service levels and/or calculation methods throughout the Implement, and Run & Improve stages.

To reinforce the collaborative approach of service providers and the service integrator, an automated dashboard of service level achievement was created and made widely visible. The front-page visualizations were dedicated to the end-to-end service levels. Then, a drill-down view was available to reveal individual service provider successes and failures for improvement purposes.

An additional dashboard of exploratory visualizations was built for a smaller audience to enable the interrogation and investigation of data.

Findings from this dashboard highlighted several improvement opportunities to record quality, service level measurement and service delivery practices throughout the Implement stage.

In most cases, it is not the individual measurement at a point in time that is relevant, but the measurement when compared against a specific target or when trended over time. A service provider may have reached its targets every month, but if the trend is downward, corrective action may still be needed.

3.5.2.5 Types of measurement

SIAM requires different types of measurement to reflect the characteristics of a multi-provider ecosystem. Unless otherwise stated, for all the following sections, the service integrator is responsible for measuring the service providers (both internal and external), and the customer organization's retained capabilities are responsible for measuring the service integrator. Self-measurement is also a good idea for all parties.

Audit compliance

The purpose of compliance is to provide assurance to the customer organization by validating that the service integrator and service providers are adhering to integrated standards, controls, policies and processes (see section **3.1.5 Governance model**). Regular, independent audit and verification determines a service provider's compliance, and trends should be tracked over time, for example, common breaches of process.

Audit and verification will normally be separate to ongoing performance management within the SIAM model. As part of the segregation of roles, it may be useful to provide audit and verification through a different function than the performance management activities (see section **2.3.9 Segregation of duties**).

Participation in the structural elements

The health of SIAM structural elements can be measured by tracking attendance of the service providers at relevant sessions, and any trends of non-attendance. Additional measures could be put in place to track behaviors within process forums, working groups and boards, based on breaches of meeting charter etiquette guidelines.

Strategic objectives

Defining and measuring against strategic objectives helps to govern the performance of the service integrator and service providers by measuring their ability to deliver value in line with the customer organization's strategy.

Relationship quality

The quality of a relationship with a stakeholder can be tracked and improved by measuring whether they are engaging/being engaged as agreed (for example, monthly, weekly or annually).

A downward trend in relationship quality for a key stakeholder could indicate that not enough time and effort is being spent on relationship building. It could also indicate that the stakeholder has been incorrectly matched with an accountability and the match needs to be changed. The inherent sensitivity of this information means that it should be stored in a secure location according to the information security policy.

Service maturity

Service maturity needs to be measured so that improvements can be identified and implemented to improve the experience of the customer organization and service users.

A number of frameworks are available to assess service maturity. Collaboration with the customer organization to identify what 'good looks like' will generate a number of criteria that demonstrate high levels of maturity.

Examples may include:

- High levels of automation within a service
- Lower total cost of ownership (TCO) for the service
- High percentage of incidents resolved at first point of contact (FPOC) or via self-service

- High customer satisfaction (CSAT) survey results for the service
- High availability and reliability of the service, a low number of incidents/problems or a long time between incidents
- A low number of reassigned ('bounced') incidents

Measures driving undesirable behaviors

A SIAM model had a service level for incident referral. The service level allowed each service provider five minutes to determine if it was responsible for resolving an incident referred by the service desk, and if not, return it. The intention was to stop a provider taking too long before allocating the incident to the correct provider.

What happened in reality was that if an incident was not resolved in less than five minutes, every service provider passed it back to the service desk at four minutes 59 seconds – allowing them some 'free' time to investigate it, outside of the fixed time service level.

This undesirable behavior was addressed by introducing a balancing service level that measured incidents incorrectly referred back to the service desk.

Process integration

Measuring process integration will help to identify any bottlenecks and inefficiencies within process and toolset integrations, and integrations between specific processes. This should include measuring the extent of integration and automation between the service integrator and the service providers, in order to identify any gaps with specific processes or providers.

Examples may include:

- Measuring the average lifecycle time from reporting an incident to permanent resolution, giving an overall picture of the consumer experience with the service provided.
- Measuring the effect of routing of incidents to the incorrect service provider, generating unnecessary reassignments and potential service provider conflicts.
- Measuring the utilization and accuracy of the knowledge repositories (for example, the configuration management system, knowledge library), including adding or updating information, but also retrieving or linking information by all service providers.

Process and service provider integration

Relating an incident to a configuration item improves incident matching, resulting in workarounds being applied more readily to improve first contact resolution (FCR) rates. This can then result in positive feedback or improvement opportunities.

Risks and issues by governance level

To protect the integrity of the SIAM model, it is important to measure the effectiveness of the management of risks and issues across the ecosystem. These measurements should align to the levels in the SIAM governance model, to identify if risk and issues are being escalated unnecessarily.

One measure to help gauge the effectiveness and maturity of the SIAM model is to assess the number of escalations to the service integrator or to the customer that were unnecessary or that did not follow the agreed process. Although rapid escalation is sometimes warranted, wherever possible, stakeholders within an escalation process should be afforded the opportunity to address the escalation at the appropriate level.

Individuals or service providers that continually disempower lower levels of governance, by escalating too high, are quickly identified when escalations are measured. Examples of other measures are:

- Percentage of risks updated by the next action date
- Percentage of risks escalated to the next level of governance
- Percentage of unmitigated risks

This also helps to determine which levels of governance are causing bottlenecks, and whether stakeholders are bypassing the governance model.

Improvement and innovation

Return on investment (ROI) and value on investment (VOI) are two key measures of the capability of the service integrator and service providers to provide effective improvements and innovations. Empirically measuring and reporting on ROI/VOI in a business context demonstrates the value of a SIAM ecosystem and the need to invest in improvement initiatives.

3.5.3 Technology practices

3.5.3.1 Management information system

A repository should be created and managed for all the data pertaining to a SIAM ecosystem. This can be used to store information such as meeting minutes, contractual data, templates and other artefacts. This may be referred to as the SIAM Library or Governance Library and may be facilitated by using tools that can be accessed by all parties within the ecosystem.

This repository needs to be owned by an accountable knowledge management process owner and managed and maintained by a designated knowledge manager within the customer retained capabilities or service integrator layer. The structure of any supporting tools needs to be defined and access controls managed, providing relevant permissions to ensure the integrity of any confidential data. This should be defined within the overall governance model.

3.5.3.2 Collaboration tools

Social networking tools can be useful in developing collaboration between disparate teams working in different locations and time zones. The challenge is providing tools that all members of the ecosystem can access either through widespread rollout or the use of integration technology. Cloud-based tools can address this challenge.

Digital workplace

A transition to SIAM is likely to result in an increasingly digital workplace. It follows that a holistic set of collaboration tools, platforms and environments needs to be delivered in a coherent, usable and productive way.

Work in the Discovery & Strategy stage needs to produce an environment that empowers individuals, teams, the customer organization, service providers and the service integrator to share, communicate and collaborate with little to no friction, delay or challenge. This allows all these stakeholders to perform their jobs more effectively.

Although communication between the layers is important, defining appropriate channels is necessary to encourage desired behavior. For example, it is not appropriate for the customer organization to engage directly with service providers. Commercial sensitivities also need to be considered, especially where contractual details are being discussed. Defining the kinds of items that should be discussed through each medium, process forum, board or working group is important. Collaboration tools are usually for more informal information sharing and collaboration, with more formal engagements used for contractual matters.

Encouraging collaboration

After spending 35 minutes of a one-hour governance planning meeting trying to establish which method of social media the service providers and service integrator could or wanted to use, and the differing benefits of each media product, they came up with the solution: *'Let's use the customer's SharePoint site that we all have been given access to!'*

Ensure that these discussions are handled within the initial exploration considerations before starting the operational meetings.

3.5.3.3 Process harmonization through technology

Achieving process harmonization requires the application of industry best practices, coupled with process engineering and organizational change management (OCM). Given the complexity of a SIAM model, a pragmatic approach alone is not enough to capture the depth and complexity of all operations and interactions needed for process harmonization.

Tools can support the management of details and abstract them into a more consumable format. Tools can provide benefit by capturing organizational complexities via workflow and dependency diagrams, enabling consistent application of standardized work products across the enterprise. The SIAM model and cross-functional teams (cultural and geographic) must be factored into the technology considerations in the Plan & Build stage.

CHAPTER 4: SIAM ROADMAP STAGE 3: IMPLEMENT

The objective of this roadmap stage is to manage the transition from the organization's 'as-is' current state to the 'to-be' desired future state, the new SIAM model. At the end of this stage, the new SIAM model will be in operation, although if a phased transition is chosen there may still be some service providers to onboard.

The timing for the start of the Implement stage can be influenced by events in the existing environment. For example, implementation could be triggered by:

- The end of an existing service provider's contract
- An existing service provider ceasing to trade
- Changes in organizational structure because of corporate restructuring or takeover
- Planned SIAM transition project commencement

The customer organization may have limited control over the timing of some of these events. It may need to react to them by completing as many of the Discovery & Strategy, and Plan & Build activities as possible. There will be an increased level of risk if the activities from these stages are not fully completed because of a lack of time.

4.1 Comparing implementation approaches

The approaches to defining and managing the transition to a new SIAM model must be considered, along with any specific associated factors affecting the customer organization. Additionally, thought must be given to any identified risks that could affect the transition to the new SIAM ecosystem.

There are two possible approaches to implementation:

- 'Big bang'
- Phased

4.1.1 Big bang approach

A big bang implementation is one that introduces everything at once within a specified timeframe, including the service integrator, the service providers (with new or amended contracts), new services and new ways of working. This could be a planned approach or triggered by a change necessitating rapid transition to the SIAM environment.

Benefits of the big bang approach are:
- If it is successful, the organization has its new model in place quickly and can get used to the new norm. Speed, coupled with success, is likely to create a positive user experience.
- From a customer organization point of view, it also allows the achievement of a steady state more quickly, reducing risks and likely costs.
- It provides an opportunity to make a 'clean break' from all legacy issues and undesired behavior at the same time, avoiding the complexities of managing a phased approach.

There are some risks, however. Using a big bang approach means it becomes almost impossible to see if the transition has fully captured all the requirements until it is released. Using an Agile implementation approach can mitigate this and accommodate subsequent requirements that are identified for later implementation. In many instances, an external service integrator will work on a 'fixed price' contract, so it is unlikely that there will be any additional cost to the customer organization if changes are needed.

Exposing a whole organization to a new model can cause fear and disruption. The mechanisms defined in the Plan & Build stage become important here both in managing the transition of new service providers, systems and services, and in controlling the people elements of the change.

4.1.1.1 Key considerations for a big bang approach

Adopting a big bang approach may raise concerns from a delivery perspective. There could be an increased level of risk to the integrity of the new model and to the existing services that are being delivered. Existing service levels need to be maintained, referred to as 'keeping the lights on'.

These considerations provide insight to help assess the suitability of this approach:

- Preparation time to define and agree the process framework before rollout – a key consideration of the Plan & Build stage.
- Tooling strategy and alignment between the service provider organizations.
- Time and resources available to test processes, interactions and tooling before the transition.
- Level of knowledge sharing (required and possible) and the handover of authority across the ecosystem.
- Transition period between incumbent service providers and the new service integrator.
- Transition time required by each service provider and the service integrator.
- Stability period for new ways of working to become established. Estimating an early life support (ELS) period that may need to extend or decrease. During this time, service performance targets may be reduced, or service credits put on hold.
- Alignment of contracts and deliverables.
- Alignment of governance framework.
- Business change within the customer organization environment.
- Sustaining the current service targets and commercial agreements.
- Cost.

4.1.1.2 Risks associated with a big bang approach

The risks associated with a big bang approach include:

- Potential higher risk/impact to existing services when compared to the phased approach
- Misalignment of contracts and no established collaboration clauses in place in time
- The service integrator is not able to lead as they are onboarding at the same time
- The service integrator does not have time to build authority with the service providers
- Inability to scenario test or stress test the SIAM model
- Inability to track individual performance against obligations, as all are onboarding at the same time

- Risk to the business/customer organization – the change is underestimated or not communicated, and is perceived to be badly managed
- Inability to benefit from lessons learnt for onboarding each service provider, as all service providers are onboarded at once
- Risk to inflight projects and subsequent challenges with transition of projects into service delivery
- Change management/inflight project impact – potential changes and releases could be impacted by the level of change within the environment:
  - Forward schedule of change from previous incumbent
  - Patching schedules and planned maintenance affected adversely

4.1.1.3 Indicators for big bang

There are several indicators for the suitability of a big bang approach, such as:

- A greenfield opportunity without legacy contracts or incumbent service providers
- An alignment to another significant change, such as an organizational merger
- When intermediate transition states would cause significant complexity or an inability to maintain accountability during a phased transition

---

A very large, big bang implementation

A very large transition to a SIAM environment was implemented as a big bang. The implementation was for a company with global reach that implemented more than 50 global service contracts at midnight on the same day. In many cases, this was a change of service providers. Approximately 10,000 staff were affected.

The key to the transition was that it was a transfer of accountability – the new contract responsibilities were enacted and the new service providers became accountable for the transition of the actual services. In many cases, those service providers engaged the outgoing service providers for services to reduce risk, but this was the responsibility of the new service providers to manage.

A 24/7 'war room' approach was used during the first two weeks to facilitate rapid response and collaboration of service providers – both onboarding and outgoing. This outcome-focused approach allowed the service providers to optimize locally without affecting the global approach. This was a successful transition with very little impact to the business and existing services.

4.1.2 Phased approach

A phased approach counters some of the risks associated with a big bang approach. For a phased approach, the Plan & Build stage defines a series of phases, with deliverables at the end of each phase.

A phased approach can more easily exploit Agile practices. A transition project can be worked on in sprints, with each sprint deploying extra functionality, new services and new service providers within different organizational functions. One of the most important aspects to consider with this approach is ensuring that each phased deliverable is adding value to the customer organization.

Minimum viable service

Another useful technique leveraged from Agile's minimum viable product is the deployment of a minimum viable service. This is where a definition is made of the minimum amount of functionality and supporting processes that allow the customer to use the service.

This approach facilitates rapid deployment of what is important and allows support staff to focus on a much smaller area and accelerate fixes if issues are experienced. The minimum viable service can then be expanded in subsequent iterations.

Using a phased approach means that if an element fails, addressing it is not as time consuming or as complex, and it is unlikely to affect users as severely as in a big bang approach. Typically, managing in phases is easier, as real-time feedback can be given at regular intervals on necessary adjustments. This can be fed back into the Plan & Build stage, if required, so the necessary changes can be made as part of later iterations.

This approach still has challenges, especially if phases relate to strategic service providers tasked with delivering core business functions or services. In this instance, it is not always possible to limit the potential impact on core services. It is critical that an overall end date for the phased approach is specified as part of the Plan & Build outcomes, with clear and regular communication to and from users. This allows it to be clearly understood when the SIAM model is complete, and the Run & Improve stage begins.

4.1.2.1 Key considerations for a phased approach

Compared to a big bang approach, adopting a phased approach provides a decreased level of risk, but a similar level of complexity for all layers in the SIAM environment. Risks must still be considered before transitioning to a SIAM model using a phased approach. Phases could be organized by service providers, functions, processes, locations or a combination of these. With a phased approach, it can take longer for benefits to be delivered to the customer organization, so the benefits realization planning needs to be carried out carefully to take this into account.

Key considerations include:

- Establishing the service integration layer first before onboarding service providers
- Ability to embed the SIAM model and collaborative working methodology in phases
- Phasing the introduction of service providers, and their deliverables and obligations, into the SIAM ecosystem
- Implications of managing business demand and customer satisfaction during a phased approach, which will have less impact but over a longer period

4.1.2.2 Risks associated with a phased approach

The risks associated with a phased approach include:

- Services may be affected for longer and the impact on existing services might be unacceptable.
- Intermediate/transition states can create complexity and confusion.

- The customer organization experiences change fatigue, as the period of change lasts for a much longer time.
- Maintaining momentum over a longer period.
- Extended transitions can increase cost, particularly if the schedule is not managed effectively and delays cascade into subsequent phases.
- Impact on relationships with the incumbent service providers, especially if they are losing the contract but are not transitioning out until a later phase. This may drive subversive behavior, which has a higher probability of causing disruption.
- Risk to inflight projects is actually reduced, as there is more opportunity to plan the phases around them.

4.1.2.3 Indicators for a phased approach

The existence of one or more of these factors may indicate that a phased approach is the better option:

- The customer, service integrator or service providers (or any combination of these) do not possess the capability for a single, large change.
- An appetite for a lower risk and impact transition approach.
- Immaturity in multi-service provider models, indicating a cautionary approach that offers learning opportunities.
- A desire to retain control and flexibility around the timing of the phases.
- A requirement for phased cash flow. A phased approach may support cash flow and budgets. Time-specific events, such as onboarding and offboarding service providers, can be scheduled to better fit the financial situation or cycles of the customer organization.
- Legacy contracts might make phasing necessary.

4.1.3 Real-world conditions for implementation

Additional factors need to be considered when deciding on an implementation approach, for example:

- Does the implementation cover a single site or multiple sites? Are there multiple geographic regions involved?
  - o A big bang implementation on a single site is considerably easier to manage than across multiple sites.
  - o Interdependencies between locations could dictate that a phased approach is not viable.
- Does the implementation cover a single or multiple business units/functions?
  - o If multiple, then it may be preferable to phase the implementation to reduce the risk for each unit.
- If different contracts expire at different times, then the cost of early termination should be weighed against the delayed benefits from a SIAM transition when waiting for the last contract to expire (which could be several years).

- If a phased approach is adopted, what will this mean for integration between the new and old environments during the interim period?
  - This is potentially one of the most problematic areas for phased transition projects. If the new way of working is implemented in a fragmented manner, then consideration must be given to how new and old will work together and for how long this is feasible. This can involve creating interfaces, interim arrangements and sometimes duplication, which would not be needed if all elements were introduced at once. It may also mean creating documentation, such as interim contracts, operating procedures and work instructions, which cover how the ecosystem will operate in the interim period.
- Are there any other competing business activities that need to be considered?
  - Factors such as regulatory compliance, acquisitions, new product introductions and other capital expenditure programs can influence the required timescale for the move to a SIAM environment.
  - Business events or cycles, for example, avoiding any change overlapping the September to mid-January period for retail organizations reliant on the Christmas season.
- What level of risk is acceptable?
  - The generally held view is that big bang implementations have an inherently higher level of risk, because the integrated nature will usually mean that a failure in one part can have subsequent effects elsewhere.
  - A phased approach has risks for every phase and some, like delays, can accumulate and complicate across subsequent phases.
- Cost considerations.
  - Phased implementations nearly always take longer to fully complete. This generally means more time from both incoming and outgoing service providers and the project team, and, therefore, increased costs. The additional time and cost should be balanced against some of the main arguments used to support or reject the approaches, such as the ability of the business to cope with a huge level of change happening all at once, as well as the increased risk of failure. Temporary interfaces, contract end dates, and legacy systems and services can also increase the cost of a phased approach.

4.1.4 Sustaining the current service

With any approach, it is important to ensure that there is limited disruption to the services being delivered. Even in the smoothest running transition there will be impacts to services.

When preparing for a transition, consider the following:

- Manage the expectations of the commissioning organization around:
  - The potential for reduced service-level achievements for a limited period during transition and early life support (ELS).
  - The need to undertake impact assessment of affected services, phased across the transition period, and how this affects business operations during transition.
- Work with HR and service providers to ensure required resources and capabilities are available.

- If the service desk is going to be delivered by a new service provider or the service integrator, it is often a good idea to transition this first, before other service providers. Establishing the new single point of contact provides a front door to the SIAM model and helps establish relationships.
- Agree and communicate a change freeze to minimize the risks. In some circumstances, allowances may be made, such as deployment of emergency changes or changes to services that are not related to the SIAM transition, based on risk assessment.
- Continue inflight projects but consider delaying or stopping new projects being initiated during transition to the SIAM model (see section **3.3.3.7 Accommodating inflight projects**).

Within any transition, there will be varying degrees of change. In some instances, the impact may be limited. The transition team should take an adaptive approach to its work (see section **3.3.3 Transition planning**).

4.2 How to transition to the approved SIAM model

The establishment of the service integrator and service providers is a key step in the SIAM transition. Consideration should be given to the impact of evolving relationships and team dynamics, which will be subject to change.

Ideally, all the core artefacts within the Plan & Build stage should be generated before service providers are onboarded (see section **3.1 Design the detailed SIAM model**).

These include:

- Detailed SIAM model
- Selected SIAM structure
- Process models
- Governance model
- Detailed roles and responsibilities
- Performance management and reporting framework
- Collaboration model
- Improvement framework
- Tooling strategy
- Organizational change management (OCM) approach
- Onboarding process
- Transition plans
- Plans for technology and related integration capabilities – which may include manual, semi-automated or fully automated interfaces for toolsets

To onboard a service provider, there must be the capacity and capability for that service provider to execute its responsibilities. In a phased implementation, this might be an interim state while other service providers transition in a later phase. The SIAM artefacts mentioned above will form a crucial part of an effective transition. Without them, service providers will have to lead themselves and, although likely well intentioned, this can create unclear lines of responsibility, inconsistent approaches to delivery and service degradation.

It is good practice to start the transition by engaging the service integrator first, although this is not always feasible. The benefits of doing so will be that the service integrator can provide support in subsequent transition activities:

- Establishing processes and supporting infrastructure across service providers
  - o It is a good idea here to undertake user-scenario testing (for example, test a change, run a major incident, run through a contract change)
- Commencing transition activities with new service providers and services, with the additional governance and management capability of the service integrator
  - o This will support building relationships and establishing working environments from the beginning
- Support with the impacts of disengaging service providers that are not part of the new SIAM environment
- Toolset and process alignment between all parties

How to align service providers that work in different ways

Flexi Corporation's ICT department consists of multiple service providers.

- Service provider A mainly provides infrastructure operational support to the production environment and uses ITIL processes for managing work.
- Service provider B is mainly responsible for the development of an application. Service provider B uses Scrum for the development of applications.

Despite the difference in how ITIL and Scrum work, the SIAM project team has created appropriate interfaces between them. Service provider A's change management process (ITIL based) protects the production environment, by identifying risks and preventing negative impacts.

Although every change was captured under change control, to provide flexibility to development teams, the following model was agreed between service providers A and B:

- Change control over product backlog is delegated to the product owner, with service operation teams involved to ensure operational warranty requirements.
- Change control over the sprint backlog is delegated to the product owner.
- Release to development is approved by the product owner.
- Formal release and change to production is approved by a formal change management authority with input from stakeholders, including the product owner.

Some of the role mappings are shown here:

Service provider A role	Service provider B role
Service owner	Product owner
CSI manager	Scrum master

Change authority	Scrum master and product owner

4.2.1 Operational handover

The incoming service provider and the service integrator will wish to acquire knowledge from any incumbent service provider during transition, such as:

- Service definitions
- Configuration data
- Knowledge articles
- Open incidents and incident history
- Open problems
- Patterns of failure and their causes
- Current processes, procedures and interfaces
- Assets and their configuration
- External supplier agreements and service responsibilities
- Knowledge of the business and its demand/use patterns
- Service and systems designs as input to later change
- Ongoing CSI initiatives

However, this might prove to be difficult, as:

- Incumbent service providers can stall or create delays to avoid early disengagement.
- Knowledge sharing between subject matter experts (SMEs) of different service providers is often not defined until late in the transition lifecycle (and does not necessarily include the outgoing service provider).
- Incumbent service providers may be unwilling to release staff to attend knowledge transfer sessions, as this is resource effort/cost over and above the current day job or has potential to impact delivery of the current day job.
- Travel to either customer or service provider location may be an issue.
- As incumbent employees leave, the pool of skilled, knowledgeable staff reduces as the cutover date approaches.

Incumbent service providers are contracted to provide a level of service until the contract termination and may also be contracted to carry out handover activities (see section **3.1.3.11 Exit services schedule**). Although some contracts may include a list of key personnel that can only be changed with the agreement of the customer organization, very few contracts have requirements for guaranteed levels of resource. It is quite likely that the number of knowledgeable resources in an incumbent service provider will reduce, as staff either leave to go to a different organization or to another part of the service provider.

> **Valuable resources**
>
> A service provider delivered services to a number of different government departments, under separate contracts. One of these was near termination and new service providers were being procured to be part of a SIAM model.
>
> In order to avoid losing good staff through TUPE, the incumbent service provider moved them to work on other accounts. This loss of skilled resources affected achievement of service levels, but the service provider was happy to accept this, including the service credits, because retaining the staff was more important.

Key staff in any outgoing service provider will want to secure their own personal future, which may or may not be with the new service provider, or the outgoing one. However, if staff leave to take up a role in a different organization before or during the transition, they are no longer available to assist with the transition. Assumptions about critical relationships and availability of staff may not hold true. It may be necessary to incentivize the incumbent service providers to try and retain key staff until the transition is completed.

In practice, many SMEs from the outgoing service provider respond well to engagement at a professional and personal level, but can experience frustration about how much they are able and allowed to engage, over and above the day-to-day business as usual (BAU) activities.

To mitigate poor engagement, new service providers need to be clear about how their own processes are going to work within the SIAM model. Process documents and RACI matrices need to detail precise process activities and interaction points across the ecosystem and with all stakeholders.

These process documents are required for every process within the scope of the service integrator. This is no small undertaking and the effort and skill required to detail and agree these throughout the SIAM model, for even the simplest of structures, should not be underestimated.

At the point of handover, the service status should be baselined to include:

- Progress/achievement against acceptance criteria, including any mitigations suggested to address shortfalls.
- Volumes and status of inflight work items, not just projects, but incidents, problems, service requests, etc. This ensures an accurate view of the performance of the newly deployed service. For example, if there was a backlog of incidents outstanding before handover, this is not subsequently seen as an indication of the new service provider's initial effectiveness.
- Service performance.
- Status of any outstanding/incomplete transition activities. In practice, there may be less significant activities that are uncompleted before handover, especially if the handover date is fixed. The decision may be to complete these under the new service provider rather than retain transition resources. For these activities, the new owner and action plan should be recorded for ongoing management.

4.2.2 Knowledge transfer

Knowledge transfer activities need to take place to ensure that every service is fully understood. Typically, this service information is documented in service definitions, architectural and process documents, contracts, configuration data and knowledge articles. It is likely that this knowledge is distributed throughout the outgoing service provider, as very few organizations maintain the ideal single repository.

The plan for knowledge transfer needs to identify knowledge areas, what is required, who owns it, who needs it and at what stage it is required. Careful planning and escalation of issues as they arise in the transition is required to ensure that knowledge is not lost from the outgoing service provider.

4.2.3 Early life support

It is a good idea to establish an early life support (ELS) function when transitioning to a SIAM model. This function assists with resolving incidents and issues, usually for a defined period aligned to the size and complexity of the transition. The function should include representatives from:

- The service integrator (responsible for managing ELS)
- Each of the service providers
- The customer organization's retained capabilities

ELS staff should have knowledge of the services, the end-to-end architecture and the processes. They should use the defined and agreed processes, not create something different. In effect, they act as an integrated team covering incident management, problem management, release management and change management. Where possible, they should be co-located, at least shortly after transition, as this aids collaboration.

The decision to stand down ELS should be formal and recorded. The point when ELS disbands will be defined based on criteria, possibly including:

- Consistent achievement of service levels over a defined period
- Agreed level of customer satisfaction
- No high-priority incidents received in the past period
- No high-priority problems outstanding

These criteria should be measured throughout the ELS period. In these initial stages, the service integrator needs to apply close (operational) governance to ensure that all new service providers (including the service integrator itself) are complying with processes and agreements. This will require thorough auditing and reporting (focused on compliance), as well as communication and OCM to 'get stakeholders on board'. Formal (for example, audits) and informal (for example, regular communication) approaches are both needed and should be balanced carefully.

If the implementation is phased, then ELS may be in place for some time as the phases complete. It may be paused and then reinstated as phases are rolled out, especially if there are gaps between phases. The participants may also change, so one ELS team may stand down on one day and another step up the next.

4.3 Apply ongoing organizational change management

Organizational change management OCM is introduced in section **3.2** in the Plan & Build stage. Within the Implement stage, OCM needs to manage the impact of the transition to the new SIAM environment. During this stage, there could be changes to processes, job roles, organizational structures and technology. It is the individuals from the service provider, the service integrator and the customer organization's retained capabilities who must ultimately change how they undertake their roles. If these individuals are unsuccessful in their personal transitions and if they do not embrace and learn a new way of working, the transition to the SIAM model will fail.

OCM guides how to prepare, equip and support individuals to successfully adopt change. When implementing a SIAM model, OCM will address the people aspects of the transition by:

- Supporting the integration of new service providers and the decoupling of incumbent service providers that are not going to be part of the ecosystem
- Creating clarity and understanding of the new governance framework
  - o Supporting changes to working practice, including coaching individuals through changes to process or team dynamics
- Guiding communications, sponsorship and training:
  - o Conducting awareness campaigns throughout the organization
  - o Communicating with and preparing stakeholders for the change
  - o Measuring the effectiveness of communications and organizational change activities
- Helping to support the diagnosis of struggling functions, service providers or process

4.3.1 Staff morale and motivation

The driving force for collaboration is motivation. For collaboration to succeed, each party involved needs to feel that they gain something or that they are providing a meaningful interaction towards a valuable result.

Transfer of services from one service provider to another affects the livelihood and careers of staff. Changes occur around them over which they may have little or no influence. This can be deeply unsettling and may cause staff to reappraise their career choices and future with an organization.

Maintaining morale

The management of staff within an outgoing service provider and the maintenance of morale towards the end of a contract, is particularly challenging for even the best of managers.

There can be a struggle to maintain concentration on business as usual (BAU) obligations and service levels, and at the same time there are demands on the existing service staff to participate in knowledge transfer and other requirements from the incoming service provider. Strong management support will be required.

Senior staff often concentrate on the commercial transition and give less priority to their workforce. This can deliver surprises when staff suddenly resign. The loss of key staff can materially affect the knowledge retained within the service environment and increase the risks to service continuity and a successful transition.

Loss of key staff may, however, also act as a positive enabler during a transition. The rest of the team pull together and create a stronger team, with different team members having the chance to step up and lead.

4.3.1.1 Positive stakeholder engagement

How someone goes about the work they do, the people they work with and the position and direction of the business they work for, affects how engaged they are. Fully engaged individuals will positively promote the work they do and the interests and reputation of the business they work for.

Improved engagement starts with understanding the objectives and commitments of the business, identified through vision, mission and values statements. All individuals in the SIAM model need to know what the business of the customer organization is, how it operates, its strategic direction and their contribution and role in this endeavor. For service providers and any external service integrator, this understanding is in addition to their knowledge of the goals of the organization they are employed by. The vision for SIAM is created in the Discovery & Strategy stage, led by the customer organization and the service integrator.

Building and maintaining positive stakeholder engagement comes through:

- Regular, positive communication and regular meetings, which reinforce the value of the work people do (for example, 'town hall' meetings or one-to-one sessions)
- Asking for feedback and listening to what all team members have to say
- Providing people with the tools and resources they need to do their job effectively
- Supporting people with professional development and training (such as ongoing professional development plans, communities of practice or mentors)
- Caring for staff welfare and well-being
- Recognizing and rewarding extra effort and outstanding results
- Encouraging fun and humor in the workplace

Within a SIAM ecosystem, these activities need to happen across the SIAM layers, within each layer and across the functions and structural elements.

Transitioning to a SIAM environment presents a unique set of circumstances, some of which include:

- Aligning multiple organizations and their management cultures
- Dealing with 'subcultures' within individual organizations
- Working within legislation when transferring employment
- Working with multiple preferred methods of communication across multiple organizations

4.3.1.2 Displaced staff

Moving to a SIAM model often involves considerable changes to staffing, especially if the move involves using additional external service providers and a service integrator and/or offboarding existing internal or external service providers. There are issues associated with staff displacement that should be considered, including:

- **Motivation and morale:** in situations where the transfer will take some time, there are likely to be issues in motivating staff to operate at the required level as the transfer commences, when they are about to become redundant.
- **Knowledge transfer:** incumbent staff may be unwilling to share information or support staff within the new SIAM model.
- **Staff quality:** some people would rather stay with their old employer or wait to be dismissed and trigger a redundancy package. This can result in issues where less motivated staff remain and motivated staff leave.

The customer organization must ensure that any relevant legislative implications are factored into the transition plan and any commercial and contractual agreements.

Working with HR

Open communication and working with HR and any staff side (trade union) representatives is incredibly important. Breaching employment law or breaking existing working practice agreements can add time and expense to any initiative and create reputational damage. Ignoring such basics also runs the risk of undermining trust in the initiative. Ensure that plans for change have the full backing of all stakeholder groups.

Impact of staff displacement

The way an employer deals with staff displacement has powerful implications for those retained. A botched program can leave a workforce feeling shaken and demoralized. This is not good news for fostering the necessary inter-team, inter-provider requirements to work as a cohesive and collaborative unit.

Keeping staff on side and feeling positive about the organization has important implications for future performance. The best approach is to ensure that all parties are well informed. The timing of the announcement is crucial. Waiting until Friday afternoon and asking staff to clear their desk is likely to be not only unlawful but also hugely inconsiderate.

Move to an outsourced SIAM model

A British multinational insurance company is headquartered in London. It has approximately 33 million customers across 16 countries.

The organization took over an automobile breakdown cover company, based in the UK, offering global roadside assistance. This company's IT had been wholly insourced. Given the global nature of the mother organization and its complex web of providers, it decided that, to support this new business, it would move to a SIAM model using the services of a renowned service integrator. This external organization was to take over responsibilities for the service desk and the service integration role, and was based in India.

A staff consultation period commenced, with HR, legal and union representatives. These stakeholders were engaged many months before the planned transition. No staff chose to take roles in India and it was agreed that staff would be displaced over an 18-month period, to allow for a smooth handover.

Several of the incumbent staff went to India to support new staff with the transition during a period of early life support. The transition went well, largely because of the investment in good staff, robust training and the gradual testing and transition elements.

The incumbent, displaced staff were generously rewarded for their service during that critical period. It is difficult to motivate staff who are losing their jobs. In this instance, the additional experience they gained as well as the opportunity to work overseas was coupled with a financial reward.

Although incumbent staff were unhappy to be losing their jobs, they also acknowledged the company's care in handling the situation. They spoke favorably about their experiences with their previous employer, helping it to maintain a good market reputation within the UK.

Do not assume that an employee will always be difficult after being made redundant. Personal pride means this is often not the case and some employees welcome a new opportunity or a financial settlement. Although some additional measures may need to be considered to protect data, knowledge, intellectual property (IP), etc., in most cases such methods are not required.

Staff who remain at an organization after their colleagues have been laid off may experience feelings likened to bereavement, and it can leave them demoralized, anxious and desperate to find a new job. Many good staff are lost in this way, which can affect ongoing operations.

Fear, uncertainty and distrust

Following the move to a SIAM model, a London-based homelessness charity's issues were evident, in particular with regard to retained staff.

The charity had onboarded a new external service integrator into the organization, a move that some staff felt had not been handled as well as it could have been. The organization had made some redundancies to accommodate the new service integrator, as well as changing some of the incumbent staff roles within the retained capabilities. This included altering some of their terms and conditions.

To resolve the issues, a workshop was held with the charity's staff and the service integrator, with the aim of helping them to get to know each other's roles better, as well as enabling them to feel that they could challenge each other in a constructive way. Staff were also encouraged to generate new ideas for improving the charity's services to clients and were empowered to present these suggestions to management in a way that explained the pros and cons.

Workshop attendees reported that, following the event, the team was 'more open', had a 'greater clarity of purpose', and that the opportunity to air their views had left staff feeling more confident and better motivated.

One of the outputs of the workshops was a change in the strategy for motivating internal staff and service providers by creating an award, for which potential recipients were nominated by the charity's users.

4.3.2 Communicating with and preparing stakeholders

Transition to a SIAM environment is likely to be led by the customer organization and its project management office (PMO), with assistance from the service integrator once it is appointed and operational. It will benefit from a dedicated communications team to ensure continuity (see section **2.2.3.2 Project roles**). Each service provider should also have its own communications team, responsible for communication within its own organization.

Who is involved and how they are involved is a critical success factor (CSF). As initial messages are drafted and target audiences are defined, it is important to be aware of the needs of different stakeholders. Some staff will be critical in driving the initiative forward, some will need to understand a change in how services are delivered to them and some will need to be prepared for new ways of working. Use analytical techniques such as stakeholder maps to identify the level of interest and influence of each stakeholder (see section **2.6.5 Stakeholders**).

From that analysis, four special types of stakeholders are key to organizational change management (OCM):

- Executive sponsor
- Project manager
- Core communications team
- Communications advisory team

Executive sponsorship is a key component of successful adoption of the SIAM model. This needs to be someone with enough influence and interest to deliver messages with credibility and authority. Many SIAM transitions fail because they do not secure executive support for communication early enough.

The project manager is the enabler and communicator who can connect with all the stakeholder groups. They have an important role in ensuring the cadence and sequencing of communication and messaging. For example, it can be very important in some cultures that senior managers are aware of a message or announcement before the rest of the organization.

The *core communications team* is the group that does most of the work preparing communication for the initiative. This should be a small group, comprising people who provide specialist communication. This team will need to have marketing skills and experience in OCM methods such as awareness, desire, knowledge, ability and reinforcement (ADKAR) (see section **3.2 Organizational change management approach**).

The *communications advisory team* is an inclusive group that supports the core communications team. It may be an existing management team, board or forum (with a representative of each of the stakeholders – service integrator, customer retained capabilities and service providers), or a smaller team of self-selected individuals from key groups (for example, HR, trade unions, staff associations). This team is engaged for input and may help test communication ideas for the project team.

Internal service provider teams need special care

Internal service provider teams often need additional care as a stakeholder group. They may expect to be treated differently to external teams. A SIAM transition will challenge this perception because they will have the same accountabilities as an external service provider. History shows this can be a difficult organizational change and care should be taken to manage this class of stakeholder.

4.3.3 Conducting awareness campaigns

During the Implement stage, awareness campaigns can help communicate important messages about the new environment to appropriate stakeholders. SIAM transition awareness campaigns typically have five main components:

1. Establish the objectives of the campaign
2. Establish a target audience
3. Define awareness campaign strategies
4. Be certain of the message
5. Measure the success of the campaign

Establish the objectives of the campaign

Establishing the objectives for the awareness campaign requires an understanding of what needs to be achieved. What are the messages to be delivered and mediums to be used? What attitudes or opinions need to be achieved? These objectives should be specific, measurable, agreed, relevant and time-bound (SMART) so that efforts can be evaluated later. It is not enough to go out and simply raise awareness.

Awareness campaign

A UK logistics organization set up an initial three-month awareness campaign with the sole intent of defining and beginning to adopt a 'single customer culture' as it moved from a single outsourcing partner to a multi-service provider SIAM model.

Establish a target audience

Having a specific target audience ensures the message is directed at relevant groups of people. In certain instances, this will be self-selecting. For example, a campaign encouraging service providers to adopt a process will undoubtedly be aimed at those using the specific process (see section **4.3.2 Communicating with and preparing stakeholders**).

Assessing the target audience identifies different groups to contact. For example, raising awareness of new service desk processes in a SIAM environment will require subdividing of the target audience. There may be a different focus for people working for the service integrator, those in the customer organization's retained capabilities, service provider staff or indeed the users of the service desk.

Define awareness campaign strategies

The ability to formulate appropriate campaign strategies depends on knowledge of the target audience. For example, the ideal locations for town hall meetings are in areas easy to reach and local to the employees involved.

Champions

Consider identifying 'champions' within the customer organization to promote SIAM, assist with the awareness campaign, and help with training and knowledge transfer.

Be certain of the message

Knowing all the facts about the message to be delivered is a precursor to a successful campaign. It is important that the team delivering the awareness campaign is prepared and has the relevant knowledge to back up the message at all levels. For example, local team leads must understand the messages being sent from the center and be briefed in how to deal with feedback and questions.

Concentrate communication on addressing:

- What is going to change or happen and when?
- What is 'in it' for the stakeholder group concerned?
- What are the implications for the stakeholder?
- How does this support the desired outcomes of the SIAM transition?
- Are there short-term risks and workarounds that may affect the stakeholder or where cooperation is required?

Prepare the briefing

An organization that moved to a SIAM model made effective use of carefully planned briefing sessions. Direct managers communicated to staff so that they had an opportunity to ask questions without fear of reprisal or concern about speaking in a larger corporate forum.

All 'briefing managers' undertook training regarding message delivery. They were also provided with pre-scripted answers they could refer to regarding anticipated questions.

A senior manager was made available after each session so staff could ask additional questions, or the briefing manager could do so on their behalf and communicate back to them.

All questions asked were collated to extend the list of FAQs, which was made publicly available on the corporate intranet.

Measure the success of the campaign

The last step is evaluation, to understand whether the communication has reached the audience (and reached them effectively). If communication efforts are a success, it is important to recognize what the audience will learn, what they will understand and believe, and how they will act.

Identify what is important to measure. There are several metrics that might be considered to assess the impact of a campaign:

- **Activity metrics**: these metrics can provide a better understanding of behavioral change. For example, activity metrics can help to evaluate process adoption.
  - Consider whether there is a discernible change in attitude or behavior among the target audience members.
- **Engagement metrics**: these metrics can help to demonstrate the effect the communications messages are having on those who hear them. Engagement metrics are a measurement of when and how others engage.

- Consider conducting a survey to understand if the message and mediums suited the audience.
- **Impact metrics**: these metrics are all about what is really trying to be achieved. Impact metrics help to measure the behaviors and attitudes that have been shifted and the wrongs righted.
    - Consider reviewing process outcomes for success. For example, look for a positive change in attendance of the structural elements.

It is important to have a monitoring system to track results. A myriad of online tools exist for tracking traffic and engagement levels, but it takes a human being to monitor both offline and online conversations and messaging to see if the new way of working is becoming the norm.

Facilitate preparation

A service integrator was going through a major change affecting how it was structured. As most of the staff were based in one location, the communications team held a weekly face-to-face communication and feedback session.

Attendance soon declined after the first session, down to an average of 30 percent. The feedback was that staff had other meetings back-to-back with this session in other buildings and some worked from home.

The format of the session was changed to online, and attendance rose to 90 percent and remained at that level throughout the project.

Finally, consider how to 'close the loop' as part of communication. How are feedback and comments collected and addressed? If staff do not see responses to their feedback, they will be less likely to engage in future campaigns.

Regardless of the measurements used, it is important to benchmark the results gathered against the milestones identified for the transition plan and trend the results over time. In general, communication activities do not produce results immediately. This is especially true where the transition has taken a phased approach, as the overall benefit of the SIAM transition may not be measurable until all phases have been completed.

4.3.4 Measure effectiveness

Measuring the people side of change is a necessary success indicator of the transition to a SIAM environment. However, the evaluation of organizational change management (OCM) is an extremely complex, difficult and highly political process in many SIAM transitions, especially where numerous subcultures exist.

Three major areas should be considered when seeking to understand the balanced effect of communication and OCM activity:

- Individual employee metrics
- Project performance metrics
- OCM activity metrics

These should be defined for every organization in the SIAM ecosystem, to understand their areas of weakness. They should be analyzed by service provider and by process area/function, and actions taken to address any issues. Experience shows that issues are more likely to be expected with groups that do not take OCM seriously.

The individual employee metrics below are commonly used when demonstrating OCM effectiveness. Many of these measures identify where employees are in the change process and how they are progressing:

- Process adoption metrics
- Usage and utilization reports
- Compliance and adherence reports
- Employee engagement, buy-in and participation measures
- Employee feedback
- Issue, compliance and error logs
- Service desk calls and requests for support
- Awareness and understanding of the change
- Observations of behavioral change
- Employee readiness assessment results

Project performance metrics can also help to measure ongoing OCM where it is possible to isolate the effects of specific actions. The metrics below expose how important it is to have delivered a thorough planning phase at the beginning of the SIAM initiative to identify these measures:

- Performance improvements
- Progress and adherence to plan
- Business and change readiness
- Project KPI measurements
- Benefits realization and return on investment (ROI)
- Adherence to timelines
- Speed of execution

Finally, it is possible to track the OCM activities in isolation. Regardless of the type of change, all structured OCM initiatives involve these activities, making these metrics useful for any project:

- Tracking of OCM activities conducted according to transition plan
- Training participation and attendance numbers
- Training tests and effectiveness measures
- Communication delivery
- Communication effectiveness

4.3.5 SIAM social network

A SIAM environment should redefine traditional collaboration by including a social angle. This solution should not just be confined to the traditional and formal interactions (for example, email or online meetings), but also extend to social networking functionality such as micro-blogging, wikis, profiles, tagging and feeds. A strong social network capacity within the SIAM environment can help to foster openness.

SIAM social network tools include:

- **Collaboration:** allows the customer organization, service integrator and service providers to share and distribute information with others, including documents, email, instant messages, online meetings, video conferencing, webinars, screen sharing, etc.
- **Content:** standard format for documents, video and images that can be used throughout the SIAM social network
- **Communication:** smart utilization of search engines, bookmarking, news feeds, profiles, commenting, photo sharing, favorites and ratings

---

Social media guidance

Guidance should be given to staff on the appropriate use of social channels, of social norms, acceptable language and content – together with a robust information security policy and outline of the obligations of all stakeholders.

In a case of dispute, any recorded communications may be produced in court. This may prove embarrassing at best and end in legal action or contract termination at worst.

Different service providers will often be remotely located, sometimes in different countries. In these situations, the value of social tools in promoting and fostering informal networks and channels can be significant. Where staff are co-located, regardless of the service provider organization, the value of social network tools is likely to be smaller. However, full co-location is rare within a SIAM environment.

4.4 Applicable SIAM practices

Practice definition

"The actual application or use of an idea, belief or method, as opposed to theories relating to it."[32]

Within the SIAM Foundation BoK, four types of practice are described:

1. People practices
2. Process practices
3. Measurement practices
4. Technology practices

These practice areas address governance, management, integration, assurance and coordination across the layers, and need to be considered when designing, transitioning or operating a SIAM model. This section looks at each of these practice areas and provides specific, practical considerations within the Implement stage. Note that the people and process practices will be combined and referred to as 'capability'.

---

[32] *www.lexico.com/definition/practice.*

4.4.1 Capability (people/process) considerations

The implementation of and transition to a SIAM environment is a major change for most organizations. Elsewhere, the necessity of organizational change management (OCM) has been discussed (see section **4.3 Apply ongoing organizational change management**). OCM capabilities need to be present in almost all layers and participants of the Implement stage.

Specifically, the following capabilities need to be employed:

- The **customer organization's retained capabilities** need to remain accountable for maintaining the governance and direction of the overall implementation. New contracts come into place, while old ones are terminated or in some cases modified. This will require commercial and legal capabilities to act on escalations from the service integrator if contractual obligations are not being met.

- The **service integrator** is accountable for managing its own activities, plus all activities of the service providers in the Implement stage. Not only will it have to put its own operations in place, but it also needs to manage, assure and, where appropriate, integrate those of the service providers that are moving into the SIAM model. Another practice that is specifically applicable during the Implement stage is the management of cross-functional teams, which might also be virtual (see section **3.2.3 Virtual and cross-functional teams**).

- The **service providers'** role in the Implement stage includes aligning their processes and people to the SIAM structural elements, becoming familiar with the terms within collaboration agreements, considering process interfaces, understanding individual contract requirements and targets, as well as end-to-end service requirements.

Particularly during the initial 'settling in' period, the attitude of 'fix first, argue later' needs to be adopted to overcome any initial issues and successfully establish the SIAM environment.

Fix first, DISCUSS later

The concept of 'fix first, argue later' is not unique to a SIAM environment, but was defined from a SIAM perspective within the **SIAM Foundation BoK**.

Although often literally applied in the real world, the concept is not suggesting that there should be an argument. Rather, it suggests that the initial focus needs to be on **fixing** the issue at hand. Discussions on who was responsible for the issue ('argue' in the phrase) will take place after resolution (or at least once there is a reduced urgency).

Furthermore, this is not intended to be an argument or finger-pointing exercise, more of a 'lessons learned'. Ideally, perhaps the phrase is better served by replacing the word 'argue' with 'discuss', 'settle', 'negotiate', 'debate', 'analyze' or 'resolve', which have more positive connotations and involve the next steps, for example, a post-incident review.

Some other people and process practices to consider in this stage include:

- Process (and tool) integration testing
- Establishment of the structural elements (boards, process forums and working groups)

- Training (of people, in the process, measurement and technology practices)

4.4.1.1 Process integration testing

A good way to test processes is to use a technique known as 'conference room pilots' or 'walkthrough tests'. In these, end-to-end tests are designed to walk through a process (and any related processes) and its interfaces between providers and the service integrator.

For example, a test could be:

- Service desk receives an incident from a user
- Service desk triages the incident and passes it to service provider A for further investigation
- As this is happening, more calls are received and the priority is escalated
- Service provider A requests assistance from network service provider B
- The root cause is identified and an emergency change is raised
- The change is approved and the fix is implemented

These tests should be carried out using operational tools and processes, but away from the production environment. The aim is to test that the documentation is correct, that everyone knows what to do and that there are no gaps or overlaps.

Conference room pilot

During a test of resolving incidents through a chain of three service providers, when it came to escalation from one service provider to another, the service desk agent said 'So, I know I should call service provider B, but what is its telephone number?'.

The test was marked as 'failed' and was rerun once the service desk confirmed that the service provider's telephone numbers were now recorded in the service management tool and supporting artefacts.

These tests can also be used as acceptance testing for the customer organization, as they demonstrate that all parties know their responsibilities, have the required capabilities and can work together effectively.

4.4.1.2 Establishing the structural elements

The structural elements are defined in the governance model. Although some of the governance boards should already be in place (see section **2.3.7 SIAM governance roles**), the remaining structural elements are formally established during the Implement stage, including the process forums. These forums should have met at least once before the first transition activities and should have reviewed their terms of reference and had them signed off.

Within the Plan & Build stage, each structural element will have defined its attendees and who can now be assigned, along with its purpose, meeting type (for instance, face to face, online, telephone conference) and meeting frequency. Regular face-to-face meetings can be impractical in many SIAM ecosystems because of geographical spread. In this case, it is good practice to have an initial face-to-face meeting and then once a year, with the intermediate meetings using online/telephone conferences.

4.4.1.3 Training

Delivering an effective training program is an important element of a transition to a SIAM model. Training all the staff required in the time available can be a challenge. Train too early and they could forget most of what they need to know. Train too late and they may not be able to absorb all of the information.

Bringing together a variety of cultures and learning styles in a SIAM ecosystem will expose any weaknesses, and this usually requires traditional training methods to be enhanced. Organizations must aim to create training programs that are fit for purpose, based on the needs of those requiring training. They must allow all stakeholders to maximize their learning potential while gearing their experience towards their particular interests.

There are five important characteristics of effective training programs that must be kept in mind to ensure best results on an individual and organizational level. These characteristics prepare participants for the types of behaviors needed in a successful SIAM model, and will increase the effectiveness of the training program:

1. **Personalize**: allow for the organization of information into modules and packages for different types of people and their individual requirements.
2. **Challenge based**: allow for participants to share and comment on challenges within the training scenarios, working together to solve issues and share advice. Discovering answers in such an environment is proven to embed learning much more quickly than traditional methods.
3. **Collaborative**: allow participants to work across teams and enhance their learning experience. Teamwork is crucial to any SIAM ecosystem and training programs should foster this type of work ethic.
4. **Multi-topic:** ensure that training programs incorporate various areas of knowledge, not just a single area of expertise.
5. **Networked:** ensure that as many people as possible can use the resources you provide, anytime and anywhere.

Training that ignores culture is unlikely to be effective. It is important to ensure that there is sufficient time made available to individuals to complete their training, perhaps by offering to cover their role temporarily. In other organizations, training is accommodated in the background without fuss.

4.4.2 Measurement practices

During the Implement stage, the SIAM model is put in place. Towards the end of the stage, there is a 'transition readiness' phase where the Run situation is approximated during final testing, without it being business as usual (BAU) just yet. This involves:

- Enabling end-to-end service measurement
- Running end-to-end service measurement

In these early stages of running the environment and measuring performance, the initial focus should be on compliance, rather than efficiency or even effectiveness. It is also a time where there may be some adjustment of the measurements and the reporting framework.

Other measurements to consider are 'progress metrics' of the Implement stage and initial metrics to measure the benefits of the SIAM transition, baselined against the benefits planned at the Strategy stage. These might include percentage measures of:

- SIAM roles/structural elements filled
- Service providers onboarded
- Service providers' internal training completed
- Services mapped across service providers (service grouping model completed)
- Processes implemented (and the maturity of these processes, and/or percentage of interfaces completed)
- Services operating to expected levels (at least as high as pre-transition)
- Contractual obligations met
- Measurements against acceptance criteria

4.4.3 Technology practices

The tooling strategy outlines the requirements for a toolset or toolsets to support the SIAM ecosystem. It will include functional and non-functional requirements, the processes that need to be supported, standards for interfacing to the toolset(s) and a roadmap for future development.

In the Plan & Build stage, the integration requirements helped to create the tooling strategy (see section **3.1.9 Tooling strategy**), which in this stage needs to be implemented, including the following steps:

- The service integrator implements or integrates the toolsets as per the strategy
- The service integrator refines the policies and standards for data exchange in conjunction with the service providers
- Service providers either integrate their tool, implement the common one if a single tool is to be used or implement 'swivel chair' procedures
- Each service provider tests its own tools, to ensure that it can provide the required outputs
- The service integrator tests its tools
- The service integrator runs end-to-end tests of the tools, ensuring that all interfaces operate as designed

Note that temporary tools for the migration of the outgoing service provider's data may be required, potentially including a manual or 'swivel chair' approach, as well as temporary reporting accessing different data sources.

The Implement stage is the first time the success or the limitations of the tooling strategy will be experienced, such as:

- Ineffective legacy tools
- Non-compliant service provider
- Lack of architecture
- Unidentified faults

Together with the measurements (and process) practices of the previous sections, the completion and success of the technology implementation needs to be scrutinized.

CHAPTER 5: SIAM ROADMAP STAGE 4: RUN & IMPROVE

The Run & Improve roadmap stage will usually commence when the Implement stage is completed, although often some aspects of the transition will still be in Implement when the stage begins. If the chosen implementation approach is 'phased', the Run & Improve stage will take on elements of delivery in an incremental way as each phase, service, process or service provider exits the Implement stage (see section **4.1.2 Phased approach**).

The inputs to this stage have been designed during the Discovery & Strategy, and Plan & Build stages, and then deployed during the Implement stage and include:

- SIAM model
- Process models
- Governance framework, including the structural elements
- Performance management and reporting framework
- Collaboration model for service providers
- Tooling strategy
- Ongoing improvement framework

5.1 Operate governance structural elements

The structural elements are intended to provide stability, support and governance of the SIAM ecosystem, enabling collaboration activities, easing smooth running and focusing on continual improvement.

Governance boards have an important role in the control of the overall SIAM ecosystem. During the Discovery & Strategy stage, the high-level governance framework is created (see section **2.3 Establish a SIAM governance framework**). In Implement, this is transferred to the live environment. In Run & Improve, the governance boards, process forums and working groups perform their roles.

A success factor for a SIAM ecosystem is that a customer organization is able to define, establish and continually adapt the service integration governance. For example, **Figure 23** illustrates a SIAM governance model where COBIT is used to establish the service integration governance (and structural) elements.

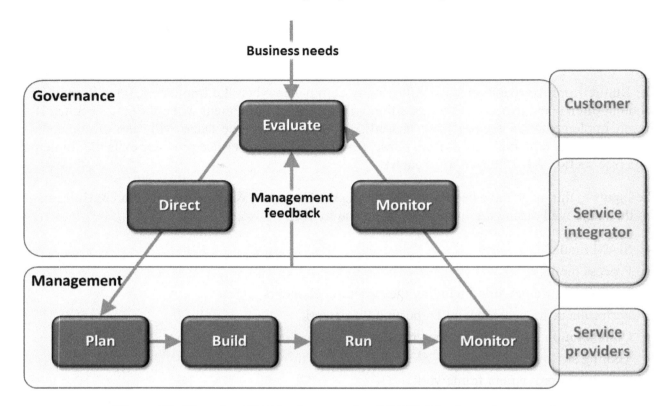

Figure 23: Mapping SIAM roles onto the COBIT 5 business framework

Governance activities are carried out at strategic, tactical and operational levels through governance boards. Boards are decision-making bodies and provide the required level of governance in a SIAM environment. In complex environments with many different service providers, more boards might be created to address specific areas. However, in less complex environments with fewer service providers, fewer boards may be more appropriate.

For example, there may be fewer boards within an ecosystem that has smaller service providers with fewer operational staff who can contribute. To reduce the overhead on these service providers, it might be necessary to reduce the number of meetings by combining boards or process forums, or reducing the meeting frequency (see section **2.3.7.4 Governance boards**).

Using COBIT

The Bank of Blue is a multinational bank and financial services organization, headquartered in London, UK. It provides the following services:

- Retail banking
- Corporate and investment banking
- Credit card solutions
- Home loans services

The IT division of Bank of Blue has recently implemented COBIT for enterprise IT management and IT governance. As a first step, the IT manager decided to implement control objectives for the management of service requests and incidents. As the service integration team in Bank of Blue IT is responsible for end-to-end management of service requests and incidents across multiple service

---

providers, the IT manager requested the service integration manager to propose how an appropriate governance structure could be implemented.

Each COBIT governance and management domain and process is mapped to related guidance. This guidance describes the standards and/or frameworks with detailed reference to which section of a standard document or specific framework the guide is relevant.

At an IT governance and management layer, the COBIT management process 'Manage Service Requests and Incidents' was used to define the required governance and enterprise needs from this management process.

Below is an example from the COBIT publication *Governance of Enterprise IT based on COBIT 5*:

- Process Number: DSS02 (it is a management process in COBIT)
- Process Name: Manage Service Request and Incidents
- Area: Management
- Domain: Deliver, Service and Support
- Process Description: Provide timely and effective response to user requests and resolution of all types of incidents. Restore normal service, record and fulfil user requests, and record, investigate, diagnose, escalate and resolve incidents
- Process Purpose Statement: Achieve increased productivity and minimize disruptions through quick resolutions of user queries and incidents
- RACI Chart:
  - DSS02.01: Define incident and service request classification schemes
  - DSS02.02: Record, classify and prioritize requests and incidents
  - DSS02.03: Verify, approve and fulfil service requests
  - DSS02.04: Investigate, diagnose and allocate incidents
  - DSS02.05: Resolve and recover from incidents
  - DSS02.06: Close service requests and incidents
  - DSS02.07: Track status and produce reports

The service integrator defined the integrated service request and incident management process to fulfil the needs of IT governance, as defined by the COBIT management process guidelines.

The service integrator also defined the principles and policies for this process and let the service providers define their own work instructions and procedures to support the process.

5.1.1 Strategic governance: Executive boards

Executive boards provide governance and oversight at the most senior level.

The attendees for these boards are senior staff with accountability for their organization's role in the SIAM ecosystem.

The central tenet of governance is transparency across decision-making processes. In a SIAM ecosystem, the executive boards need to demonstrate this with regard to any decisions made, from the initial investment in the SIAM model (doing the right thing), right through to delivering the desired benefits during the SIAM transformation (doing things right).

These boards hold the service providers and the service integrator to account for their performance and should:

- Adopt agreed standards and policies
- Set priorities and approve resource allocation
- Make decisions that are acted upon
- Reward as well as censure
- Adjudicate escalated issues

In addition to the executive board attended by all service providers, each service provider has an individual board with the customer and the service integrator. This allows a service provider to discuss commercial performance and sensitive issues with an appropriate audience. If the service integrator is also an external organization, it may need to leave these individual boards for specific commercial discussions.

The Executive Steering Board is the executive board responsible for setting the SIAM vision and directing the role of SIAM governance within the larger context of IT governance. This board should also include attendees from principal parts of the business to represent consumers of the services (see section **2.3.7.5 Executive Steering Board**).

The early focus of this board is on achieving the SIAM implementation. In the Run & Improve stage, it changes its focus to the operation of the SIAM model. This can be a challenge for some attendees who are more at home with project delivery. There is sometimes a change of attendees at this stage, from senior project team members to senior service delivery representatives.

There are likely to be unfinished activities from the previous stages. If a phased transition approach has been adopted, the next phases will still be in the planning, building or implementation stages. The board should maintain a focus on these as well as on the live services. This is normally achieved through having separate agenda items.

Some organizations choose to keep two separate boards until all the phases have completed – one focusing on live services and the other on projects that are not yet live. Although this can seem to be a good idea and can work well, it can also lead to conflicts and tension between the two boards and if not carefully managed can result in gaps, overlaps and confusion.

Project boards that never end

A large SIAM transformation project involving multiple service providers had an executive project board that had been established for some time. The executive project board had been in place before the design and implementation of the SIAM model and its associated governance boards. The focus of the executive project board was on project delivery. They had strong relationships with all the service providers. Once the services were live, this board continued to meet and started to discuss live service issues.

For some time, there was conflict and confusion between this board and the Executive Steering Board established under the SIAM model. Both boards considered that they were responsible for live service issues, but there were still project activities to govern as well as the services to be developed and enhanced. The service providers preferred to attend the project board as they had built good relationships with the attendees.

The resolution was to terminate the executive project board and transfer its responsibilities to the Executive Steering Board.

There is a significant risk that the executive boards try to deal with too much detail, including items that should sit with the tactical and operational governance boards. This is particularly the case where the same people sit on the different boards, as can happen with smaller SIAM ecosystems.

In this situation, it is important to establish:

- Clear terms of reference for each board
- Agenda items that detail what should be discussed
- Strong chairing capabilities
- Defined procedures for escalating and devolving items between the different board levels

5.1.2 *Tactical governance boards*

Tactical boards sit between the strategic and operational boards. They undertake preparation activities in readiness for the strategic board, and can be used to carry out discussions before meeting with the customer organization, or another stakeholder, at an executive level. They should also be used to identify items for escalation to the strategic board, and act as a point of escalation for operational boards.

These boards are not typically attended by the customer, and will be chaired by the service integrator, acting as the customer's agent. In some instances, such as the early days when the service integrator role is being established, representation from the customer organization's retained capabilities may be needed to provide initial support and to reaffirm the integrator's role.

Service review board

A service review board is an example of a tactical board. Its activities may include:

- Ensuring alignment of SIAM medium-term strategies with IT governance and the vision as given by strategic boards
- Optimizing the design, delivery, operations and sourcing of services
- Providing recommendations to the strategic boards for change in contracts, service providers or financials
- Conducting an annual review of service provider performance, service improvements and the service portfolio
- Reviewing the key potential risks from the operational boards

5.1.3 Operational governance boards

The main operational board convenes to discuss service performance at a lower level than either the strategic or tactical boards.

It will review service performance and act as an escalation point for all other operational boards and process forums. For example, it may authorize budget or resources to carry out improvement activities identified in a process forum that exceeds its approval limit.

Other operational boards will be scheduled as required to support decision making. The most common example of this is the integrated change advisory board. To provide a clear view of the operational environment and support the operational boards, it is commonplace to use a visual management tool to display service performance information.

Using visual tools to assist operational boards

The service integrator at the Bank of Blue implemented and ran a Kanban board (in this context 'board' means a tool for visualization and not a governance body). This used Lean systems thinking to create representations of service status and issues on physical or electronic whiteboards.

The Kanban board was used by an operational governance board at two levels:

- The level one operational governance board was attended by team leaders for respective service provider teams at a component level
- The leadership team used the level two board, which covered an integrated view of service and any escalations, bottlenecks or concerns from level one

The Bank of Blue found that this approach provided the following advantages:

- The status and outcomes were transparent to all, providing a plan and supporting dialogue
- It provided an indicator for team leaders and supervisors to react and if necessary stop to initiate a countermeasure plan
- It facilitated discussions about performance across teams at all levels

---

Figure 24 shows an example of a simple Kanban board that can support visual management (see section **5.7.2 Measurement practices** for more on visual management).

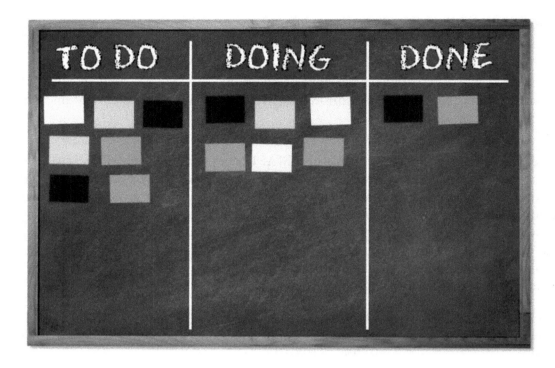

Figure 24: Kanban board

5.2 Process forums and working groups

At an operational level, working groups and process forums all help to establish relationships and encourage communication between service providers and the service integrator. These working groups and process forums are part of the structural elements of the SIAM ecosystem, spanning the SIAM layers.

There are many possible process forums and working groups that can be implemented. Decisions regarding what is required are considered during the Plan & Build stage, but the value of these must be evaluated on an ongoing basis. The service integrator must work to balance the requirement to bring teams together against the impact on service delivery. It is necessary to ensure that they do not create a challenge where the overhead of participation may be too much, as this will negate the value.

Example of a traditional working group: Major problem working group

A major problem is any problem where the severity is such that it is deemed necessary to perform urgent problem analysis on the issue with the intent to identify the root cause. Within a SIAM model, this would be carried out via a working group. The scope of a major problem analysis may include people, process, measurement, environment, technology and material.

Techniques such as Kepner-Tregoe problem analysis can be used to facilitate a major problem review or root cause analysis (RCA). This technique can be used when bringing together a group of subject matter experts (SMEs) within a working group.

The scope of a major problem working group is typically:

- Major incident investigation where RCA is required to restore services

---

- Recurring incidents leading to major problem analysis
- High-priority problem analysis to avoid possible high-priority incidents

Since many service providers have experience of traditional service management approaches, such as ITIL and COBIT, they are often comfortable with the working group format and can engage successfully in this environment.

Structural elements such as process forums in a SIAM ecosystem are typically aligned to a specific process or practice. Members work together on proactive development, innovations and improvements. It is acceptable to combine process forums and working groups where there is a case to do so. For example, a problem management forum may exist that has within its scope an action to convene a working group when a major problem is identified or indeed a problem record backlog needs to be acted on. Similarly, process forums can be amalgamated, for example, an integrated change and release forum. In each instance of adaption, it is necessary to ensure understanding of the scope and intent of the group, and undertake ongoing value measurement.

When operating in a relatively stable environment, it may make sense to introduce a multi-layer structure for control and governance, aligned to a more traditional and formal structure. In an evolving or change-driven environment, flexibility and less formal structures for control and governance may be best.

It is important to understand that structural elements are not limited to using service management frameworks such as ITIL, or standards such as ISO/IEC 20000 only. With the adoption of Agile methodologies into the service management discipline, the structural elements may use practices based on Agile and Scrum. For example, Agile retrospectives could be considered as a process forum under the end-to-end continual improvement elements within a SIAM ecosystem. A retrospective can be used to discuss what could be changed that might make it more productive next time.

It is common to use the following questions during retrospectives:

- What went well?
- What did not go so well?
- What should we do differently next time?

Example of Agile-based SIAM structural elements

The Clearwater organization is a service provider of water, wastewater and drainage services. The company employs more than 5,000 people and manages an asset base of $25 billion.

Clearwater primarily provides the following services:

1. Main water supply schemes
2. Wastewater systems
3. Irrigation
4. Drainage

Clearwater was running a significant project to add additional services and improved service levels to its portfolio of mains water supply services to end customers, and wanted the IT department to provide a faster service to Clearwater as a business.

The IT division started using Agile principles to support the business. The initiative was focused on improving mains water supply services by enhancing the automated billing system. The IT division was having issues in maintaining consistency with this Agile practice across the many service providers. Because of these ongoing inconsistencies, the head of IT asked the service integration manager to introduce Agile retrospectives as a new process forum for the service providers of the billing system improvement initiative.

Initially, the service integrator felt that Agile rituals could not be considered as structural elements, and they should be managed outside of the SIAM ecosystem. The Scrum Master explained that the activities undertaken within the process forums and governance boards would be appropriate and requested a trial.

The service integration manager tried the initiative, which demonstrated some success where service providers unfamiliar with Agile practices recognized its value and implemented it within their individual service provider process activities.

Commercial matters should be excluded from operational process forums. Top layers of governance are the appropriate settings for discussing contracts. This operational level should be reserved for discussing operations and improvement activities.

Daily or weekly standups

Daily standups are an Agile technique, now commonly used outside the software development environment. Standups are where members of a team meet every day for a quick status update, ensuring that all the main parties are aware of current issues, major events planned for the day and to raise any concerns. The idea is to stand up, to encourage keeping the meeting short (no more than 30 minutes). They are often held next to boards that provide visual supporting information. Where teams are not co-located, as is often the case in a SIAM ecosystem, standups may be held using collaboration tools.

Daily standups are associated with frameworks such as Scrum and Kanban. These two Agile methodologies are often used interchangeably, but there are differences between them.

Scrum

Scrum is a framework used to organize work into small, manageable pieces that can be completed by a cross-functional team collaborating within a prescribed time period (called a sprint, generally two to four weeks long). The aim is to plan, organize, manage and optimize this process.

Kanban

Kanban is also a tool used to organize work for the sake of efficiency. Like Scrum, Kanban encourages work to be broken down into manageable chunks, such as backlog (the to-do list) and work in progress (WIP). The work can be visualized as it progresses through the workflow using a Kanban board.

Where Scrum limits the amount of time allowed to complete an amount of work (by means of sprints), Kanban limits the amount of work allowed in any one condition: only so many tasks can be ongoing, only so many can be on the to-do list.

Whichever approach is used, it is recommended that ground rules are established for standup meetings. Three valuable questions to be answered at a standup are:

- What did you do yesterday?
- What will you do today?
- Are there any impediments in your way?

Daily, or regular standups (sometimes referred to as 'daily prayer' meetings) can be used to ensure that service teams have the right focus. Scrum or Kanban standups can be very useful techniques for working groups as they can be used to tackle a specific objective, task or issue.

Example of daily standups

A standup meeting can also be used within a specific service management process activity, such as problem or incident management.

At the Bank of Blue, a problem involving three different service providers was taking a long time to investigate, as the providers were in different geographical locations. A daily standup was held using a collaboration tool. This was led by the service integrator, to share progress and keep track of all investigation actions.

5.3 Ongoing performance management and improvement

Metrics help a business determine whether its goals are being achieved, but are effective only if they have been carefully chosen to represent progress towards objectives.

The model for monitoring and measuring service performance should have been initially considered during the definition of the governance framework in the Discovery & Strategy stage (see section **2.3.14 Monitoring and measuring service performance**), and refined during the Plan & Build stage (see sections **3.1.7 Performance management and reporting framework**), before being implemented. In the Run & Improve stage, the performance of services and the service providers is actively measured.

The framework will continue to develop over the initial period (Run), as experience highlights areas where improvement is necessary or possible (Improve), as business objectives and supporting metrics evolve and knowledge is acquired. The service integrator should own this framework. Once in place, a periodic review undertaken by the service integrator will ensure that the correct elements are measured to assess the ongoing value of the SIAM model.

The performance of all services and processes should be measured and monitored against key performance indicators (KPIs) and defined service level targets. The measurements should be both qualitative and quantitative, and show both point-in-time performance and longer-term trends.

Although it is important that each service provider has measurable service targets to work towards, they need to form part of the end-to-end performance management and reporting framework. This will, in turn, provide evidence of demonstrable achievement of service objectives, business benefits and value.

If there is no clear definition and communication of value or end-to-end metrics, service providers may focus only on their own performance and not see the big picture. Commitment to contractual requirements for managing performance is defined in the Plan & Build stage, and the service integrator is responsible for engaging with service providers to ensure their obligations are met.

5.3.1 Key performance indicator mapping and service-level reporting

Measurements are used to create meaningful and understandable reports for various audiences across the SIAM ecosystem. They provide visibility of performance issues, and support trend analysis to provide early warning of possible failures or potential delivery issues.

In some cases, a service provider might identify that it is likely to miss a target in one area, possibly because it is focusing resources in another area following previous issues. It is a good idea to make the service integrator aware of this as soon as it is identified, as the service integrator could help the service provider to prioritize when there is a conflict between individual targets and end-to-end service targets.

Reports should be used not only to measure service achievement and value but also to identify opportunities for improvement and innovation. Routine service improvement activities should include review and management of actions arising from the information and review of report relevance. Within a SIAM ecosystem, reports also need to include feedback about how the service is perceived by consumers, referred to as qualitative reporting (see section **3.5.2.4 Choosing the right measurements**).

The complexity of a SIAM ecosystem can make the production of reports a considerable overhead. Although the reporting provides value at various levels, it should not be allowed to become all-consuming.

The following types of report are useful:

- **Service provider-focused reporting**: this describes how each individual service provider is performing against its commercial service level targets and KPIs. It describes the overall commercial picture, highlighting where measures have been achieved and describing where failures have occurred and why they happened.
- **Service-focused reporting:** this focuses on the performance of the services provided, in terms of service level agreement (SLA) performance and specific targets, for example, the processing of incidents, problems and changes.
- **Business/customer-focused reporting**: this focuses on the performance of the SIAM ecosystem in terms of end-to-end services, and is perhaps the most useful in providing the customer organization with insight into the quality of services being provided, especially when expressed in business terms. The number of major incidents is one measure, but if you can translate that into lost production it has more meaning and can result in better support for corrective action.

Example KPI mapping

A service integrator wanted to compare the performance of each service provider's change management procedures. Each provider used a different internal procedure as part of the end-to-end change management process.

The change management forum was asked to create a set of consistent KPIs. The change managers from the service providers and the service integrator developed a simple set of KPIs that were easy for each provider to measure, but which gave a good indication of performance. At the end of each month they sent their KPI results to the service integrator, who then collated and shared them with all service providers. This drove competition and hence improvement.

The KPIs were:

- Percentage of emergency changes
- Percentage of late presented changes (the target was five days' notice for non-emergencies)
- Percentage of changes rejected as incomplete by the service integrator
- Percentage of failed changes

An aggregate score was created, with varying weightings on each KPI.

All of these were trended over months and presented as part of the service integrator's consolidated service report.

Shared KPIs are useful when it is necessary to compare the performance of different service providers in a specific process area. There needs to be a consistent definition on what is included in the KPI. They also need to be carefully designed to ensure that any comparison is valid.

5.3.1.1 Service-focused reporting

This requires metrics that focus on the entire service offering, requiring shared, dependent and related service levels that track the collaborative delivery of the services aligned to business outcomes. Metrics that offer no value to the business, or that cannot be understood in business terms, are likely to be ignored and risk damaging the relationship between the service integrator and the customer organization.

Measuring the performance of a service requires:

- A focus on the value provided to the customer organization
- Measuring the end-to-end service performance

Using (near) real-time data when reporting on service-focused KPIs can provide many benefits, as depicted in **figure 25**.

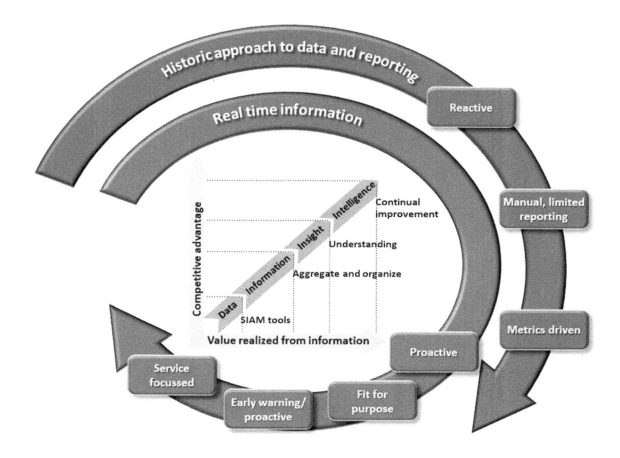

Figure 25: Reporting using real-time information

In order to facilitate this approach, it is important to:

- Implement any changes necessary to be able to capture the required service measurements within the agreed tools
- Where possible, develop an automated reporting capability
- Monitor service level compliance
- Work with service providers to improve performance

It is important that the service integrator works with the customer organization's contract management function to ensure any changes are incorporated and reflected in SLAs and contracts.

SLAs and KPIs should be a part of defined service commitments. It is often useful to run a pilot exercise as these measurements are being established or refined so that there is clarity on expectations, confirmation that service targets are achievable and that performance baselines can be established.

5.3.1.2 KPI aggregation

The role of KPI aggregation is to demonstrate both end-to-end service delivery and the effectiveness of the service integration layer. For a KPI reporting framework to be valuable, metrics and targets should be developed jointly with the service providers and support the customer organization's business outcomes. A performance management and reporting forum can be useful in the development and maintenance of aligned service metrics and the sharing of measurement best practices.

Aggregated KPIs should:

- Focus on results of services and demonstrate the impact of service delivery on business outcomes, not the output of service providers.
- Provide both qualitative and quantitative measures to give a balanced view.
- As much as possible, be objective, to minimize subjective interpretation.
- Be **s**pecific, **m**easurable, **a**greed, **r**elevant and **t**ime bound (SMART).
- Be appropriate and the total number of KPIs should be limited across the organization (such as approximately three per goal).
- Have boundaries. If a service provider does not control all the parts of service performance, then that service provider cannot be held responsible for failure to meet targets.

KPI aggregation across a complex multi-service provider model may be challenging unless the concept of shared KPIs is well understood and communicated across all layers within the SIAM ecosystem (see section **3.1.7 Performance management and reporting framework**).

5.3.1.3 Reporting tool

Ideally, reporting should be generated from the service toolset, which should act as the single or central authoritative source of service data from across the SIAM ecosystem. This 'single source of truth' reduces the need for manual data manipulation and provides a trusted basis for all reporting. Some of the data may need to come from each service provider's tool, to be consolidated in the service integrator's tool, depending on the tooling strategy (see section **3.1.9 Tooling strategy**).

If the service integrator's tool cannot meet the reporting requirements, it may be appropriate to supplement this with a specialist reporting tool with sophisticated data analytics capabilities. Although 'simple' spreadsheets are still used even in the largest of organizations, they are not as reliable and can be subject to errors. However, they are a good proving ground for new measures before complex reports are built in a tool.

Analytics capabilities will need to be built gradually to ensure they are sustainable, and to ensure there is strong governance in place to manage ongoing reporting requirements. Failure to do this could result in the service providers and service integrator meeting every reporting requirement presented to them, but failing to provide business value through the provision of insightful and useful reports (leading to the 'watermelon effect').

5.3.1.4 Reconciliation between different providers

There will be occasions where a service provider fails to meet a target because of circumstances outside its control, typically when the failure was because of another service provider. One useful approach to deal with this is the 'excusing cause'.

In this approach, the affected service provider provides the service integrator with the full information for why the failure occurred, including details of which service provider was the cause, supporting metrics and timelines. The service integrator then considers the request to excuse the failure, often consulting the 'causing' service provider.

The service integrator has the following options:

- Reject the request.
- Accept the request, allow the affected service provider to resubmit its performance report with this failure removed, and ensure that the 'causing' service provider has taken it into account in its performance measure.
- Accept the request, allow the affected service provider to resubmit its performance report with this failure removed, but leave the issue as 'unresolved' because it cannot prove that the 'causing' service provider was responsible for the issue.

This may require invocation of a dispute resolution process (see section **5.3.4.3 Dispute resolution**). The service integrator needs to have defined delegated authority to prevent decisions needing to be escalated to the customer organization's retained capabilities.

5.3.2 Differentiation between provider and integrator performance

The contract or agreement for the service integrator will usually have broad, aggregate-level targets to track end-to-end service performance. As these targets are not necessarily directly actionable metrics, they may not provide unfiltered visibility into service performance.

The performance measurement in place in the SIAM model needs to track:

- Individual service provider performance
- End-to-end service performance
- The performance of the service integrator and how it is fulfilling its role

This requires different types of targets.

5.3.2.1 Measuring the service integrator

It is often challenging to measure the value of the service integrator. On the surface, it may seem simple: if the end-to-end service is running well, the service providers are performing and the customer organization is happy, then things must be going well.

Measuring the service integrator's value is far more difficult than measuring an individual service provider, so a degree of innovation is needed. The measures need to be experiential and behavioral, for example:

- Analysis of governance activities undertaken, for example:
  - Reduction in service credits
  - Alignment of service performance to the customer organization's strategic objectives
  - Adherence to legislative or regulatory obligations
- Effectiveness of the structural elements, for example:
  - Percentage attendance at governance meetings of all appropriate service providers
  - Reduction in disputes between service providers
  - Reduction in disputes with the service integrator
  - Reduction in escalations to the customer organization
  - Effectiveness of governance meetings in planning and risk reduction terms, including how many risks have been identified and how many require mitigation/action plans?

- Process maturity and integration measures, for example:
  - o Achievement of end-to-end service targets
  - o Percentage of incidents/requests allocated to the appropriate service provider
  - o Reduction in the number of incidents passed between service providers
- Ability to coordinate the demand, scheduling and delivery cycles of the customer organization, and feed this into capacity and availability plans
- Collaboration – subjective measure of the performance of the service integrator in driving collaboration across the ecosystem
- Improvement – driving through improvements, running successful service improvement plans and coordinating actions across service providers within the ecosystem, for example:
  - o Increasing usage of the shared knowledge management repository
  - o Percentage of suppliers involved in collaborative improvement initiatives
  - o Percentage of improvement initiatives that have achieved a quantified positive business impact
- Innovation – demonstrable evidence of genuine service innovation, as opposed to improvement

5.3.2.2 Dealing with incomplete or non-standard data

In any SIAM ecosystem there may be some service providers that are part of the SIAM model but have not agreed to tailor their performance reports to align with the SIAM model reporting standards. The impact of this can include:

- Inability to obtain reports
- Incomplete measurements and data
- Irregular reports
- Misaligned reporting periods and deadlines
- Different calculation methods

This situation typically occurs when the following service provider types are part of the SIAM ecosystem:

- Commodity or standardized service providers that provide the same service to all of their customers with minimum customization, and will therefore not tailor any reporting for a specific contract
- Large service providers that provide the same reports to all of their customers
- Specialist service providers, where the service integrator's standard requirements are not aligned with the characteristics of the service delivered
- Small service providers, where the cost of meeting the reporting requirements is disproportionally high when compared to the value of the contract

For reporting purposes, non-compliant service providers leave the service integrator with one of the following choices:

- Exclude the measurements from end-to-end reports. This can lead to an inaccurate picture of service performance where these services are an essential part of end-to-end delivery, such as hosting services.

- Take the data from the service provider and do the calculations again. This can be a challenge if the service provider uses different reporting intervals or does not provide the base data.
- Take its own measurements of the performance and availability of the services. This is likely to require specialized tooling to capture the data, but will provide the most accurate information.

5.3.2.3 Imposing service credits

Service credits are pre-specified financial compensation that the customer organization may become entitled to when a service level or target is not achieved (see section **3.1.3.6 Service credits**).

The challenge within a SIAM ecosystem arises in determining when it is fair to apply these, since many service providers are likely to be contributing to the end-to-end service. The service integrator, acting as the customer organization's agent and provided with delegated authority, often has to consider the appropriateness of levying service credits.

When applying service credits, it is important for the service integrator to approach the calculation by considering how they might affect the other remedies that would otherwise be available under the contract or common law. Unless the contract is carefully drafted, an approach that attempts to impose service credits on a 'no-service-no-pay' basis can lead to situations where the pre-specified service credits become the exclusive remedy for serious performance failures.

The application of service credits can only be done fairly within an effective measurement framework. This is constructed in the Plan & Build stage (see section **3.1.7 Performance management and reporting framework**). In practice, when disputes over performance arise, the monitoring of service levels may be the most complete record of performance under the contract. The service integrator may wish to show that there have been breaches and, indeed, that the failure to meet the service levels is a symptom of bigger failings. However, the service provider may also wish to rely on the service-level reporting to show that it did everything that the parties regarded as being important.

This is where the complexities lie in a SIAM ecosystem. Often, service providers are requested to forgo meeting their commercial targets to achieve a benefit for the end-to-end service. In this case, the service integrator must apply credits considerately, in a fair and equitable manner. If this is not done well, relationships will soon break down, making collaborative working unlikely.

5.3.3 Evolving ways of working

Ways of working are established in the Plan & Build stage, but will need to evolve in the Run & Improve stage. The role of the service integrator is to create an appropriate environment through:

- Information sharing in an open manner
- Transparent decision making
- 'Win-win', 'can do' attitude
- Impartial and equitable imposition of service credits
- Recognizing that if an approach benefits all parties, it is most likely to succeed in:
  - o Promoting trust and reliance on each other
  - o Supporting collaborative working
  - o Discouraging protective behavior, while recognizing commercial realities
  - o Sharing of concerns and ideas for issue resolution so that no party feels unduly threatened or compromised

5.3.3.1 Evolving through operational management

The operational management role of the service integrator in a SIAM ecosystem is multi-faceted and typically involves several elements.

Real-time business as usual (BAU) management deals with the monitoring of incident queues, escalations regarding compromised or breached service levels and coordination of the resolution of major incidents and problems. The service integrator must ensure BAU management is successful, without doing the BAU activities. The leadership abilities and cross-functional skills of individuals involved in the ecosystem operation are key to success. Understanding the effectiveness of BAU management, or where it needs to be more effective, can focus attention on evolving improved capability where it is most critical.

Periodic meetings are necessary to review performance and provide a mechanism to ensure governance within the ecosystem. These reviews support vital collaboration activities. Reviews are an opportunity to demonstrate both the capability of the service integrator's coordination and management role, and the service provider's service delivery capabilities. Monthly service provider service reviews, cross-service provider and single provider service reviews are examples.

Service review meetings may highlight where some service providers are performing better than others and may have better ways of working to share, or conversely highlight aspects of poor performance, where service providers need to find better ways to deliver service in order to achieve the required objectives (see section **5.3.4.1 Service provider review**).

Process forums provide excellent opportunities to evaluate the overall effectiveness of the processes in operation within the SIAM ecosystem. Process forums allow the service integrator to identify operational challenges and drive continual improvement. Examples include Continual Service Improvement And Innovation Forums Or A Quality Management forum that facilitates discussion across teams and at all levels.

Similarly, the process forums themselves should be evaluated for their effectiveness and alignment to the needs of the SIAM ecosystem. Consideration should be given to the need, scope, objectives, achievements and stakeholders for these structural elements. Terms of reference should be reviewed, amended and approved by the appropriate senior staff and circulated to all participating parties to ensure ongoing agreement and value.

Collaboration remains a key attribute required to Run & Improve a SIAM ecosystem. Areas that may be investigated to assess the success of collaboration include:

- Participation in cross-organization problem solving working groups
- Membership and active participation in the SIAM structural elements, such as cross-organization process forums and meetings
- Consistent and prompt payment
- Clear and meaningful reports
- Decisive action promptly delivered when promised
- Consistency
- Delivery on obligations, supported by evidence
- Effectiveness of relationships

- Sharing of knowledge and experience for the benefit of the organization
- Openness and even-handedness in addressing issues, clarity on what both parties need to do for outcomes to be effective

---

Collaboration in the family

Managing service providers in a SIAM ecosystem has similarities to the dynamics within a family: parents (the service integrator) need to be firm with their children (the service providers), but not authoritative, instead guiding and directing them (and being fair to all children/service providers equally).

Trust here is not solely based on age and position (*'do what I say'/'because I said so'*), but also on a long-term relationship that has proven beneficial.

Only when working together ('give and take') will optimal benefits for all family members be achieved.

5.3.3.2 Driving improvement and innovation

To drive improvement and innovation, the service integrator should define methods to help stakeholders within all SIAM layers to work productively and collaboratively. The challenge in any environment is the people. People can be both the biggest contributor to and resistor of change.

These elements will provide focus to encourage improvement and innovation:

- Focus on individuals
- Focus on the team
- Customer organization and retained capabilities
- Processes
- Innovation and improvement outcomes
- Psychological climate
- Physical environment
- Organizational culture
- Economic climate/market conditions
- Geopolitical culture

Focus on individuals – the basic building block of getting things done is an individual. Organizations, departments, divisions, groups, teams, etc. are all units built from individuals. Focus on strengthening the primary building block to start pushing innovation activities forward.

Focus on the team – individuals make things happen, but in most cases, they cannot do it all by themselves. Innovation requires multiple skill sets, whether it is invention, development, funding, marketing, patenting, operations, etc. Those skill sets almost never exist in one person, so it requires several people to move it forward.

Different service providers have different skills. Focus on improving effective and collaborative team dynamics to keep the innovation engine running smoothly. Involving all layers in the creative and innovation process increases the probability that the innovation will be successful.

Customer organization and retained capabilities – even individuals in successful teams can become resistant to change. The successful innovation team of yesterday becomes the 'this is the way we've always done it' team of tomorrow. The customer organization needs to give thought to creating and sustaining enterprise-wide procedures, policies, metrics, recognition and executive level accountability to keep innovation running.

Processes – establish how to improve the processes or methods being used to drive innovation, but do so across all three levels described below:

- **The individual level:** for example, processes to enhance self-awareness, emotional intelligence and cognitive ability
- **The group level:** for example, using a structured brainstorming, ideation or creative process to support teams in creating innovative solutions
- **The enterprise level:** for example, the organizational system for idea management

The structural elements, particularly process forums, provide an ideal opportunity for this type of collaboration.

Learning to trust

Immediately after a SIAM implementation, a service integrator wanted to approve every change made by every service provider. This was because of two things: lack of trust and the change management staff in the service integrator wanting to use their experience as operational change managers.

In view of the number of service providers and high volume of changes, a change advisory board was being held twice a day, every day of the week. This continued for ten months.

Eventually, the service integrator introduced an approach where low-risk, repeatable changes that were local to a service provider could be approved by that provider. This immediately reduced the workload, enabling the board to meet twice a week.

Over time, as trust improved, the service providers were encouraged to approve their own changes, under a change management policy developed by all parties.

Several years later, the integrated change advisory board now only meets by exception.

Innovation and improvement outcomes – innovation and improvement are two different, related topics. Improvement is gradual within scope (doing the same, better), whereas innovation is a step change that could affect many parts of the business (doing the same, differently, or doing different things).

There are various perspectives of the improvement and innovation processes. To focus only on a product or outcome is to overlook services, business models, alliances, processes, channels and more. The service integrator should consider the broader picture of improvement and innovation opportunities, driven by reporting and feedback loops built into the SIAM model.

Psychological climate – reporting alone should not drive the innovation and improvement efforts. There is much to be learned about the quality of services within a SIAM ecosystem by listening to stories. User experience stories, service provider stories and customer organization stories all help provide focus on where improvements are needed.

- What is working?
- What is not working?
- What is acceptable?
- What has changed in the industry?
- What is our scope?

The SIAM ecosystem will need to evolve along with changing business and customer requirements. To drive innovation, the right amount of personal freedom (within boundaries) should be offered to all the layers of the ecosystem, offering the capacity and scope to explore new areas. Supporting an effective psychological climate is a requirement for sustained innovative output.

Physical environment – this is often an issue in SIAM environments because of the dispersed nature of the parties. There may be physical challenges in terms of separate commercial organizations operating over various geographies and time zones (see section **3.2.3 Virtual and cross-functional teams**).

This is a challenge for the service integrator to overcome. It should consider:

- Are stakeholders at the various layers able to get together easily to communicate and work?
- Do they understand their scope and boundaries?
- Are they able to make time for these activities?
- Are decision-making accountabilities clearly defined?
- Is there an appropriate space to review document prototypes/results/data?

Different people have a different concept of the ideal environment. It is imperative that the service integrator works with key stakeholders to ensure that the appropriate environments for enabling collaboration meet all parties' needs. This often requires alternative approaches, various forums and methods. Engaging all parties in defining the environments will enhance the likelihood of success.

Organizational culture – in a SIAM ecosystem there will be more than one culture evident. When onboarding service providers into the SIAM model, it is important to consider cultural alignment. This may not always be possible. Unique providers can have a unique culture and cannot (and should not) always be brought 'into line'.

Developing an understanding of the different cultures allows the service integrator to understand how best to engage with them. To gain clarity about the cultures evident within the layers, it is a good idea to look at the structural elements to understand the stories that people tell about success and failure.

- How do people discover and share how things really get done?
- Which practices are in force to work around established processes if they are not fit for purpose?
- Which processes or activities do people avoid?
- Which service providers are deemed as easy to work with, or not?

What organizational leaders say is often drowned out by what people know is really going on. It is not enough to just say innovation is important! The customer organization must provide the framework, scope and boundaries for it during the Plan & Build stage, so that the efforts can happen in the Run & Improve stage. Organizational policies, management behavior, things that are measured and executive messaging must all align to create the stories that explain the desired culture.

Economic climate/market conditions – an innovative culture is easiest to maintain when market conditions mean there is not too much fear, nor too much confidence. These are rare moments in the business cycle.

In a fast-evolving ecosystem where there is significant change, innovation can and will fade away during periods of disruption, such as service provider retirement, organizational cutbacks and restructuring activities. Service providers will usually 'play it safe' and stop making innovation and improvement suggestions when they are aware that sales are down, or that the economy is in decline. Similarly, if the customer organization announces market dominance or impressive financial figures, service providers may become complacent.

The customer organization should set the tone by setting resources aside to support innovation in both good times and bad. Paradoxically, many organizations only get radically innovative when they are in distress situations: when there is no other choice but to change things.

Geopolitical culture – this is a significant consideration in a SIAM ecosystem, especially in one that operates over many regions. Local culture elements can make a difference, such as:

- Where people were born or live
- The language they speak
- Where they work
- How they were educated

Different cultures communicate differently, see the world differently, perceive different threats and find value in different things. Every culture has strengths and weaknesses. The service integrator must consider which cultural strengths can be exploited, and which cultural impediments must be overcome. Paying attention to the habits and needs of the people in all layers of the SIAM ecosystem will support innovation.

5.3.4 Ongoing service provider management

Operational management will be successful only if supported with the ongoing ability to measure and manage each service provider. The service integrator should maintain a detailed contact matrix for each service provider, defining its individual responsibilities and accountabilities for delivery.

The service integrator should perform an ongoing evaluation of the role of each service provider. It is useful in this instance to use some form of SCMIS. Ideally, the SCMIS will be an integrated element of a more comprehensive knowledge management system.

The SCMIS should be used by the service integrator to capture ongoing records about all service providers. As well as information about their contract details, it should include:

- Details of the type of service(s) or product(s) provided
- Service relationships with other service providers (dependencies)
- Importance of the service provider's role in service delivery

- Risks
- Cost information (where available and appropriate)

Information within the SCMIS will provide a complete set of reference information for any service level, service measurement and service provider relationship management procedures and activities undertaken as part of the service integrator role.

An important ongoing element of the service integrator's role is to provide information to the customer organization about the performance and value of the various service providers within the SIAM ecosystem. Through measurement and evaluation, the service integrator should identify the position and relevance of each service provider. This information will also help the service integrator to establish the appropriate level of operational monitoring and review the schedule required.

In line with the agreed performance management framework, the service integrator should review the delivery obligations and service performance of service providers in preparation for scheduled meetings.

Any instances of underperformance or conversely of exceptional additional value should be included in the regular service reporting and fed back to the governance boards and customer organization. Underperformance by a service provider should be dealt with through appropriate corrective actions, including launching formal service improvement plans.

On a more granular level, daily standups (see section **5.2 Process forums and working groups**) could be convened by the service integrator to consider operational concerns such as support backlogs, major incidents, escalations and planning for the day.

The major success factors for successful performance and operational management are:

- Clearly defined roles and responsibilities
- Constant flow of communication within the ecosystem
- Well documented measurement framework
- Efficient measurement mechanisms
- Consistent monitoring and reviews
- Course correction mechanisms
- Ability to identify and recognize exceptions
- Ability to reward exceptional value addition
- Defined 'ways of working' that are well understood by everyone

5.3.4.1 Service provider review

Service review meetings provide an important role in assuring the service(s) delivered, as well as enabling continual service improvement and refinement to take place and be tracked formally.

Adopt a layered approach when establishing service review meetings:

- **Monthly (or fortnightly) operations meeting** – to consider:
  - o Incident status discussion based on monthly reports
  - o Areas requiring focus

    o Major escalations

    o Customer feedback

    o Action Items from previous operations meetings

- **Quarterly (or monthly) service provider meeting** – to consider:
  - Service level agreement (SLA) target performance
  - Service improvement plan
  - Improvement initiatives
  - Corrective plans for issues
  - Action items from previous meetings
  - Challenges faced
  - Feedback from the customer and/or from other ecosystem partners
- **Annual contractual reviews** – to consider:
  - Engagement with the customer organization as required
  - Consolidated SIAM scorecard (see section **2.8.2 Measurement practices**)
  - Service/service scope reviews
  - Service level reviews, aligned to the annual review of the end-to-end key performance indicators (KPIs)/SLAs
  - Strategic opportunities
  - Roadmap ahead, including any necessary changes
  - Security review to ensure there are no specific security risks within the ecosystem
  - Regulatory and compliance obligations

Note that timeframes and frequencies are indicative only and may be increased or decreased depending on specific circumstances and whether the service provider is classed as strategic, tactical, operational or commodity (see section **2.3.7.4 Governance boards**).

Minutes should be taken at all meetings and made available to relevant parties in an accessible location, with action trackers to monitor activities through to completion. Such documents can be valuable inputs to contract renewal reviews to establish eligibility for renewal or changes necessary to the service levels or commercials.

5.3.4.2 *Adding and removing service providers*

Because of the nature of a SIAM ecosystem, with multiple service providers, potentially shorter contract times and agile 'loose coupling', there will often be a need for the service integrator (and the customer organization's retained capabilities) to add and/or remove a service provider (see section **2.3.13.4 Onboarding and offboarding of service providers** and **3.3.3 Transition planning**).

Reasons to terminate a contract with a service provider include:

- Consistently failing to provide services and service levels that meet business requirements
- Services no longer align to business needs
- Finding a more cost-effective, better or more reliable service provider
- Analysis of performance or demand patterns reveals changes in the volume, transactions or service level and requirements that the incumbent is unable to satisfy (inability to scale the service)
- Natural contract end date occurs and there is no desire to renew

- Cultural misalignment
- Fraudulent actions
- Service provider ceases to trade

In all instances, it is important to check the contract first to see whether there are penalties for terminating early or indeed notice periods previously agreed. Exit clauses should have been drawn up with the initial contract, and the contract management process should be aware of such (see section **3.1.3.10 End of contract**). In instances where termination is required at speed, penalties may have to be accepted by the customer organization.

As well as the financial barriers to changing or removing providers, when there is a switch to a new service provider with different processes or systems, there are likely to be operational challenges. This may lead to disruption from new ways of working, processes, tools and loss of knowledge.

The service integrator, following the exit agreements defined, should ensure that all appropriate information and artefacts are obtained from the outgoing service provider. If possible, negotiate so that your new service provider takes responsibility for handling the changeover process with the incumbent.

Considerations during offboarding or contract change

Disengaging a service provider can be a complicated and risky business, especially if its role has been a strategic one and the services it delivers are deemed to be vital to the organization. The SIAM governance framework provides a mechanism to consider the associated risks and ensure, once identified and understood, plans are created for their mitigation or management. Guidelines should have been defined at the Plan & Build stage to deal with this activity.

The customer organization will expect continued smooth running of operations during on or offboarding, using the defined procedures created for this purpose within the Plan & Build stage. Within the Implement stage, this guidance regarding on and offboarding is used. This provides for repeatable processes with detailed quality gates to govern any service retirements and/or decommissioning activity. Significant lessons learned can be gained from these activities, which should be applied to these processes in readiness for future reuse.

The objective of 'quality gate' based transition planning and support is to ensure that all required quality and performance parameters are met, including:

- Relevant intellectual property (IP), policy, process and procedural documentation is retained as appropriate
- Knowledge transfer is undertaken
- Service continuity is retained (where appropriate)
- New operational teams are engaged and trained
- Customer acceptance is obtained

It is important to avoid alienating a service provider that is still required, for example, if it is providing other services within the ecosystem or it is likely it may return in the future.

5.3.4.3 Non-conformant service providers

In instances where service providers fail to fulfil service agreements or obligations, depending on the nature of the failure, the service integrator should take the following actions:

- Undertake a full review to establish the cause and point of failure.
- Quantify the impact of the issue on the customer organizations' business operations.
- Consider, and if appropriate and/or possible, quantify the impact of the issue on other service providers.
- Convene with appropriate stakeholders through the agreed performance management process. This may be via a board.
- Consider the application of any defined contractual remedies such as service credits, versus the application of non-contractual remedies such as improvement plans. In instances where financial penalties are applied, consider consulting with contract management.

It is important to apply contractual remedies, such as service credits, consistently across all service providers to avoid any allegations of favoring one service provider over another. It is also important to apply them consistently throughout the life of a contract, as a decision to apply them later in the contract term may cause challenges from the service provider.

Whether or not contractual remedies are applied, service failures should also be addressed using measures including review meetings and performance improvement plans. During any meetings, be sure to document expectations, the success criteria and how achievement will be measured. This should include planning for how to make the improvements, the agreed communication methods and timescales. In a worst-case scenario, this documentation could be used as a record should any contract breach and subsequent legal proceedings arise.

Ensure that the service provider's senior management is aware of the failure to meet expectations. Request that they take ownership of the agreed remedial actions within their own organization, providing support as required.

Hold regular, planned progress meetings with the service provider to assess progress of improvement activities, discuss any issues and offer support where required. It is important that these meetings are seen as an opportunity to work together on resolution, otherwise they can damage the relationships between the service integrator and the service provider.

Although the service integrator is usually responsible for managing the service agreements with the service provider, it may be necessary, for example when financial remedies are applicable, to involve the customer organization in discussions with the service provider. This is also the case when an agreement cannot be reached or when a service provider is regularly failing agreed targets, and improvement actions and/or penalties have been unsuccessful.

5.3.4.4 Dispute resolution

Dispute resolution in a SIAM ecosystem needs to be multi-level, allowing disputes to be resolved at the lowest level of escalation possible. Since the customer organization retains ownership of contracts, mechanisms need to be defined to allow the service integrator to manage most service provider and contract related issues, unless they become serious enough for the customer organization's retained capabilities to step in.

To this end, it is important to draw a distinction between performance management and relationship management, which are a service integrator's concern, and contract management, which falls into the realm of the customer organization's retained capabilities.

Dispute escalations

Often, the complainant will go straight to the top of the customer organization and the issues get blown out of proportion, whereas peer negotiation might work better. The overall culture within the SIAM model can help to reinforce appropriate escalation.

There are always disputes in contract management, but the more clarity about accountability embedded in the contracts, the easier it is to resolve them. To achieve this clarity, the strategy should begin with the end in mind, and the contracts should support the strategy (see section **3.1.3.9 Dispute management**).

It is possible for contract management to become adversarial and, in extreme cases, lead to back-and-forth reprisals. This is not only because of a difference of opinion, but also because of different perspectives between the parties. Service managers are often concerned by service quality factors, whereas commercial and financial managers may be more interested in who pays what, to whom.

Within a SIAM ecosystem, contract management should create an environment of collaboration, seeking win-win scenarios. The structure of the contract and the supporting schedules need to facilitate the ability for changes to the service arrangements, rather than a multi-year lock-in with no flexibility.

When engaging in dispute resolution, attention must be paid to:

- The law – normally, the governing jurisdiction of the contract defines which law applies
- The contract, including dispute resolution clauses
- Any precedent which may affect interpretation of clauses

5.3.4.5 Using collaboration agreements

The division of services between multiple service providers creates the requirement for service integration. The obligations on service providers to participate in coordinated end-to-end delivery may be collected in a single schedule or distributed across the contract. Ideally, these obligations are standardized for efficiency.

Where they are standardized, maintain a single document under change control. These types of documents have been called various names, such as collaboration agreements, cross-functional statements of work, engagement models or operations manuals (see section **3.1.8.3 Collaboration agreements**).

Whatever they are called, these agreements must:

- Define the roles of customer organization's retained capabilities, the service integrator and service providers
- Define the methods for communication and collaboration
- Define how to escalate and resolve operational issues
- Define how collaboration will be measured, along with incentives to increase collaboration
- Be easy to understand
- Contain as much as is necessary, and no more

Simple collaboration agreements

A large organization commissioned a SIAM ecosystem. It decided it needed a collaboration agreement, so it engaged commercial lawyers to create one, at significant cost.

The collaboration agreement that was produced contained more pages than the main contract, and was written using mostly legal language. This scared away many potential bidders.

Once services commenced, the agreement was never used, as the selected service providers understood the required outcomes and wanted to 'do the right thing'.

The document was worthless. A collaboration agreement should be a living and breathing document that provides reference to support clarity of obligation.

Although it can be difficult to strictly enforce collaboration, having a collaboration agreement helps to set the tone and define the expectations around working arrangements and engagement between the service providers and the service integrator, and with the customer organization's retained capabilities.

The collaboration agreement sets the baseline for the relationships with the service providers, but should not preclude any additional behaviors that could enhance the delivery of services. Otherwise, there is a risk that some service providers may work strictly to the agreement and go no further, which may also restrict service improvements.

Collaboration agreements provide:

- Clarity regarding the overall service outcomes and individual outputs that are sought from the service providers
- Easily understood definitions of which party is responsible for what and the mechanism that is best placed to achieve these
- Clarity regarding where standardization is to be applied (for example, selection of master tooling set) and where discretion is available (for example, each service provider's internal tooling is acceptable if it integrates and exchanges with the master)
- Service agreement schedules that will consider both end-to-end and individual service provider accountabilities
- Links to any governing artefacts, including integration and interface requirements

Success will depend largely on the support of the service providers in accepting the conditions defined within the collaboration agreements. The service integrator's role is to assure operational conformity and take action over issues that affect collaborative working.

5.3.4.6 Trust-based supplier management

In SIAM, the best outcomes are achieved when there is trust between the customer organization, the service integrator and the service providers. Trust-based supplier management is an approach that recognizes this, varying the amount of governance performed by the service integrator over service providers, depending on the level of trust in each provider. This helps to further build trust, support cooperation and allows the service integrator to allocate its management time in the most effective way.

Many organizations have historically managed suppliers using an approach that relied solely on contracts. This can lead to caution and mistrust when designing a supplier and contract management strategy for SIAM, resulting in very detailed contracts with excessive reporting requirements and penalty clauses. Such organizations can find it difficult to transition to a SIAM model that requires collaboration, cooperation and trust in order to successfully manage suppliers and contracts.

Trust-based supplier management can be used instead of, or in conjunction with, these more traditional supplier management techniques. The choice of approach will depend on the nature of the contracts and the maturity of the relationship between the service integrator and each service provider.

> **Trust in individuals or trust in organizations?**
>
> Although there is often talk about the need to build trust between organizations or teams, trust actually evolves between individuals. Trust is people based rather than contract based. Trust can exist between individuals at all levels in organizations. Trust earned at C(Chief)-level may not always translate to staff at an operational level.

Trust and goodwill are important foundations for collaboration across the SIAM ecosystem and successful interactions between all layers in a SIAM model. Therefore, trust should also be considered in the design of the wider SIAM model, including the process model, collaboration model, tooling strategy, ongoing improvement framework, and the performance management and reporting framework. This provides surety and consistency across all stakeholders (see also the **SIAM Foundation BoK** on the challenge of the level of control).

The design of the detailed SIAM model in the Plan & Build stage should consider where trust is required for successful operation. This should include the level of trust required, how to build and maintain that level, and the responsibilities for making it happen. The scope of the design should include the SIAM practices, especially the People and Process practices.

The levels of trust that exist within a SIAM ecosystem will evolve and change over time. **Figure 26** shows how trust can increase over time and is also affected by specific events.

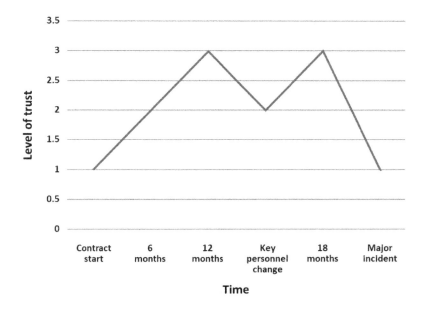

Figure 26: Levels of trust impacted by time and events

Example situations where the level of trust can be affected negatively are:

- When a new service provider is introduced to the SIAM model because its performance is not proven
- If service providers oversell their capabilities to win a contract and cannot deliver what was promised
- Following a major incident
- When key personnel change and new relationships need to be established
- When service scope is not clear and service providers feel they are punished for things that are outside of their control
- When expectations and requirements change without communication, and contractual terms and metrics no longer appropriately reflect the customer organization's needs
- Pressure for end-to-end efficiency to the benefit of the customer places a greater demand than contracted upon a service provider

Customer organizations that have decided to use an externally sourced service integrator should try to ensure that their expectations on trust align with that of the service integrator's. This will avoid issues where the customer disagrees with the service integrator's supplier management approach. As the service integrator is acting on behalf of the customer, it is important that they understand and represent the customer's underlying attitudes and behaviors towards trust.

Missing trust

Two months after the implementation of a new SIAM model, a major incident was caused by a service provider's engineer shutting down the wrong server in a data center.

The service integrator did not trust the service provider to prevent it from happening again, and insisted on new access control procedures. These required the integrator to approve every request to enter the data center.

The more controlled approach improved the level of trust between the parties. As a result, access control levels were reconsidered gradually and ways of working evolved so that the controls were gradually relinquished.

The trust management cycle

Trust management is an ongoing journey. Even in tried-and-tested relationships, influencing factors will have a direct impact on the trust relationship at a point in time.

Consider the development of trusted relationships as a journey with many stages:

- **Define trust. What is it, how is it measured?** Keep it simple, perhaps you have defined trust as confidence that you can depend on a service provider or team. This can be displayed by having multi-provider teams share tasks and feel comfortable asking for assistance from each other.
- **Understand barriers to achieving trust.** Move from service provider by service provider-based targets to an end-to-end measurement. This removes the feeling of competition and allows trust to be established (see section **5.3.1.1 Service-focused reporting**).

- **Build the foundations for a trusted relationship to occur.** Building trust takes time. It requires an awareness of the service model dynamics and the opportunity to practice the process of trusting others. Team building can be informal, such as social events or as part of the interactions through the structural elements. Evolve activities through process improvements or value-stream optimization, allowing for the development of trust and reinforcing the principles you wish to establish.
- **Support, maintain and modify the trust environment.** As the level of trust develops there will be situations that impact it. Dealing with these situations immediately and addressing any issues in relation to trust, will allow the environment to be maintained. The measurement of trust is a point in time indicator. Once you have defined what it looks like, check it often, measure and rate it, and make the ratings visible. Ensure that all layers within the SIAM model understand it and that trust may increase or decrease based on circumstance. It is not a blame game, rather a recognition that all relationships, even those that are long established, will change over time and circumstance.

Once a good understanding is gained about what management practices need to change for high-trust relationships to grow, a better understanding of the conditions that make trust-based SIAM practices necessary should be developed. Rigidly formalized methods of cooperation are replaced in favor of new principles of agile cooperation, to achieve high-performing, cohesive teams that cross organizational boundaries.

Examples of SIAM practices to build trust across the ecosystem are:

- Create a baseline 'trust level' for each service provider and identify where relationships or culture need improvement.
- Support collaboration using face-to-face interaction as well as collaboration technology and tools.
- Be inclusive of everyone and consider the social elements of the workplace, such as summer parties, project kick-offs and team building across organizations.
- Embrace group dynamic and action-based learning. Trust builds over time and the acceptance of a decision grows when the service providers are not only informed but can also discuss their recommendations and ask for justifications and explanations.

The baseline 'trust level' provides the service integrator with an understanding of, and clearly mapped, levels of trust across the ecosystem. For example:

- Low-trust relationships: intensive management is required, often including more detailed reporting and regular meetings. The focus is to confirm the service provider can get the job done to the level agreed.
- Medium-trust relationships: management becomes less intensive. The service provider has more autonomy and can focus on improving how the job is done.
- High-trust relationships: the focus moves from management and oversight to growing and maintaining a positive and healthy relationship, with a focus on shared goals and innovation.

Developing a sense of unity and cohesiveness creates the trust in which impartiality and openness take their place. This is sometimes referred to as a 'one team' attitude that encompasses all actors in the SIAM ecosystem.

Often, trust is expressed by symbols of identity, for example, logos or specific expressions that have a unique meaning to group members only. Each SIAM ecosystem will develop its own symbols, rituals and communication patterns to create identity. Trust-based supplier management can recognize and develop these expressions of identity.

Gummi bears as a reward

A SIAM team had the habit of bringing gummi bears (sweets) to work and started handing these out – as a reward – whenever someone (from outside the team) actively promoted SIAM in favor of their own way of working.

The gummi bear eventually became a symbol for SIAM and even posters in the offices had the gummi bear on it.

Challenges for growing trust include:

- A high employee attrition rate (as trust is based on a personal relationship)
- Dependence on specific persons (or service providers) and their knowledge
- Competitive situations (between service providers)
- Resistance to change (as in changing ways of working for the service provider staff or even the customer-retained organization)
- Inability to physically meet (instead relying on remote interactions)
- Cultural alignment

5.4 Audit and compliance

Most businesses are regulated in some way or another. There are compliance standards relating to industry, corporate and business governance, setting out requirements to adhere to industry practices, legal and government requirements or an internal objective to meet a certain benchmark. This results in the service providers needing to meet and maintain adherence to certain quality, audit, compliance and regulatory standards.

In a SIAM ecosystem, when the service integrator has accountability to govern service providers, responsibilities extend to the management of quality and compliance parameters. In addition, the service integrator will need to maintain audit readiness or compliance posture through record-keeping and performing preparatory internal reviews and audits.

Compliance management and audits are defined in the Plan & Build stage (see section **3.1.5 Governance model**). In the Run & Improve stage, audits should be carried out according to the schedules in the contracts, or in response to a major issue that highlights any potential non-compliance with obligations.

Audits should only be undertaken by staff with the appropriate experience and/or qualifications. In a SIAM ecosystem, this may include staff from multiple domains or service providers. Although there may be local, specific, confidential areas for individual service providers, this should not impact assurance activities, which should be achievable within an open and collaborative culture.

5.4.1.1 Driving improvements post audit/assessment

Audits are an integral part of most management systems and are usually a requirement of external standards (such as ISO 9001[33] or ISO 14001[34]) or external regulatory frameworks. Audits or assessments can identify not just non-conformities to be addressed through corrective action, but also identify issues that are either systemic or show trends that identify potential areas of weakness.

Audit activities help all stakeholders improve the organization by providing useful information to the business in pursuit of continuous improvement. Auditing should be considered a way of helping the organization identify and improve the effectiveness and efficiency of its practices in pursuit of the organization's objectives.

This is different from simply identifying compliance or otherwise. The difference between these two approaches may be regarded as the difference between facilitation and observation. Holistic audits ensure the business is 'doing things right' and also validate that 'the right things are being done'. For example, this means not just ensuring that processes are being followed, but that they are appropriate for the needs of the business and that those processes contribute to customer satisfaction, however that is defined or measured.

Value may be added during audits by triggering a discussion on best practices or making suggestions for improvement.

The objectives of audits include:

- Supporting stakeholders in the organization to deliver their goals and objectives
- Measuring the performance of business processes (efficiency, effectiveness and conformity)
- Assessing the organization's ability to meet customer requirements, and internal and external rules and regulations
- Facilitating the process of identifying and sharing best practice
- Identifying improvement opportunities, risks and non-conformities
- Supporting the adoption of external standards

An audit program that will outline the approach and frequency of audit activities, is defined within the governance framework in the Discovery & Strategy stage (see section **2.3.12 Auditing controls**) and undertaken within the Run & Improve stage based on the guidance provided within the governance framework.

5.4.1.2 Audit reports

Audit reports must be timely, supported by facts and evidence, objective, agreed between the members of the auditing team and the findings must be well documented.

There are three parts to a well-documented finding:

---

[33] For more information, visit: *www.itgovernance.co.uk/iso9001-quality-management-standards*.

[34] For more information, visit: *www.itgovernance.co.uk/iso-14001-environmental-management-systems*.

- A record of the requirements against which the finding is identified
- The finding statement itself
- The audit evidence to support audit findings

The structure of the audit report should meet the needs of the customer organization and could, for example, include:

- Scope
- An executive summary that gives an overall assessment of the health of the area being audited
- A statement of whether the area or activity reviewed conforms to the requirements placed upon it
- Any opportunities for improvement
- Any findings and areas of concern
- Areas that may be considered best practice
- Information for future audit planning
- Areas that require follow up

Targets for preparing, approving and distributing audit reports should be agreed as part of the governance framework and approved during the audit closing meeting. Where possible, audit reports should be distributed to all the stakeholders as quickly as is practicable. Audit reports that are released and distributed long after the audit has taken place may be discredited or not given priority.

5.4.1.3 Follow-up activities

The customer organization should set clearly defined timescales for the completion of actions agreed during the audit activity. For external audits, these are usually defined by the body performing the audit.

It is commonplace in many organizations that these timescales are also adopted for the equivalent internal audit finding categories. However, an alternative approach may be to consider the severity and impact of the audit finding when agreeing the target date for closure. A flexible negotiated approach (with defined boundaries) can help ensure agreement and cooperation of all stakeholders, and will increase the likelihood of successful completion.

Improvement post audit

After an audit, process A had a major non-conformity raised against it and process B had a minor non-conformity. Under the governance policy, major non-conformities had to be addressed in six weeks and minor non-conformities in 13 weeks.

Upon further investigation, process A had a six-month cycle and would not run again for another five months, while process B had a monthly cycle and was due to run again in two weeks.

With finite resources available, it would make more sense to negotiate a more effective set of deadlines that reflect this, based on resource requirements, available mitigation and risks to the business.

To help ensure that findings are completed as expected, it is also recommended that any action with a timescale of more than two months has a milestone plan agreed, with any individual milestone no longer than two months. If progress is then monitored against the milestones, the likelihood of successfully meeting the overall target date is improved.

The timely closure of agreed actions is very important to the effectiveness of the audit process. However, it is recognized that from time to time audit actions will not be addressed within the agreed timescales. This may be because of any number of factors, such as lack of resources, change of business priorities, recognition of need, etc. To manage such events, a suitable escalation process should be established that can be used in cases where agreement to address the issues cannot readily be obtained.

5.5 Risk and reward mechanisms

Risk and reward systems are designed to align the motivations of service providers to the motivations of the customer(s). Service providers must care about the end outcome and avoid self-serving behavior. Risk and reward mechanisms must be defined during the Plan & Build stage. There are a number of mechanisms that can be employed. In some cases, this is an opportunity to apply service credits or service credit earn back (see sections **3.1.3.6 Service credits** and **3.1.3.7 Incentives**).

There is also an option to become more innovative, by using mechanisms to align the service providers and the service integrator to the customer organization's goals. There are anecdotal stories demonstrating this in action, for example, where bonus payments have been tied to the revenue of the customer organization. This level of goal alignment requires maturity and high commitment to partnership and transparency. In the SIAM ecosystem, the ability of a service provider to manage the risk of a customer attaining a target can seem daunting. Service providers may be reluctant to have revenue dependent on a customer or another service provider (who may also be a competitor).

The typical approach is the use of shared targets, based on a customer key performance indicator (KPI) where attainment of the target allows financial benefit. This benefit may be graded based on the level of attainment. For example, loss of productivity of no more than ten percent because the services being provided might earn a reward, but a much larger reward occurs where there is zero loss of productivity, perhaps on a sliding scale. This may be linked to the current baseline, and to improved achievement levels over time to drive improvement and innovation.

The targets and the allocation of service credits should be managed in a dynamic way. There needs to be a mechanism at the governance level to modify the targets, either by negotiation or by a contracted change approach. Since the goal is to encourage positive behavior, locking in an arrangement over a three- or five-year contract term would be unworkable.

Attributes of the risk/reward program would include:

- Driving collaborative behavior to the desired outcomes
- Service credits and earn-back – this allows a scenario where desired, collaborative behavior is rewarded
- Shared KPIs are particularly useful in driving shared risk/reward and hence collaborative outcomes
- Monitoring the effects of the reward mechanisms in place – great care should be taken to monitor for negative behavior being encouraged by the program
- Those who take risks, reap rewards – this means asking providers to step up to higher-level outcomes, particularly with respect to grouped services

- Creating a provider performance tree (making results visible and showing who is doing well)
- Mechanisms to improve ecosystem culture and transparency between stakeholders
- The importance of knowledge management and continual training across the ecosystem

5.6 Ongoing change management

The service provider landscape is likely to change. Once the initial model has been implemented, the customer organization may choose to onboard more service providers to the model.

Transition planning and support is required not only for new service introductions, but also in cases where a service has been significantly changed. It is also important that architecture, security, delivery and other standards and policies are in place.

Harmonization between supplier management in the integration layer and contract management within the customer organization's retained capabilities, is essential to handle service provider exit and entry scenarios or in the event of a new sourcing requirement. This will create a cascading impact on the supplier management, transition planning and support, change management and release management processes.

There will be situations, such as organizational changes within the ecosystem, that will have significant impact on the people working within it. The service integrator should encourage all the ecosystem service providers to have effective knowledge management and backup plans in case of changes.

Effective operation of a SIAM ecosystem depends upon the ability of all stakeholders to understand the model and demonstrate the desired behavior. Therefore, any change in the people can have significant impact on the SIAM model, if not anticipated and mitigated early enough. As the service integrator's control on service provider staff will be limited, this is a risk that needs to be carefully mitigated.

The people perspective of the change is something that is often neglected. The guidance provided within the Plan & Build stage on organizational change management (OCM) is a useful resource here (see section **3.2 Organizational change management approach**). Another key factor to make ongoing change management effective and efficient is the focus on process integration. A change, no matter how small, will impact several elements within the ecosystem. If processes (and technology and people) are not integrated, absorbing the positive or negative impact of any change may result in significant imbalance in the overall model.

Managing a SIAM environment requires early detection of any areas where 'siloed' ways of working exist and dealing with them directly. One way to address this is having detailed lessons learned sessions soon after any change in the ecosystem. This allows for the development of an understanding of what went well and what went wrong. The lessons learned should not be superficial, but must address the layers, people, process and technology.

5.7 Applicable SIAM practices

> **Practice definition**
>
> *"The actual application or use of an idea, belief or method, as opposed to theories relating to it."*[35]

Within the SIAM Foundation BoK, four types of practice are described:

1. People practices
2. Process practices
3. Measurement practices
4. Technology practices

These practice areas address governance, management, integration, assurance and coordination across the layers, and need to be considered when designing, transitioning or operating a SIAM model. This section looks at each of these practice areas and provides specific, practical considerations within the Run & Improve stage. Note that the people and process practices will be combined and referred to as 'capability'.

5.7.1 Capability (people/process) considerations

The Run & Improve stage supports the operational delivery of the SIAM ecosystem in an incremental way, as each phase, service, process or service provider exits the Implement stage.

Initially, it will be necessary to ensure that knowledge levels and process capabilities are sufficient and mature. Often, immediately after implementation, knowledge levels and process capability maturity are the minimum required to take on services and processes. In the early stages, they will not always be proven under stress, and execution can be immature. As the model matures, the requirement will be to ensure that the capability of people and processes is optimized based on changing customer needs.

SIAM is a combination of people, processes and tools. These components need to work together effectively for a SIAM environment to run smoothly.

The following activities support the Run & Improve stage within a SIAM environment.

5.7.1.1 Ongoing capability assessment

The customer organization will have defined the expectations for people capabilities within both the service integrator and service provider layers. They will relate to the standards required to support performance and relevance (see section **3.1.6 Detailed roles and responsibilities**). Within the Run & Improve stage, the service integrator will provide assurance against these standards by managing the capability framework.

---

[35] *www.lexico.com/definition/practice.*

Each service provider will maintain its own framework and systems for assessing the effectiveness of its people. Examples include the skills of teams or functions, such as project management, service management and specialist IT staff, and the evolving digital systems they design, deliver and support.

5.7.1.2 Skills mapping

The role of the service integrator is like the captain of a ship. It needs to translate the direction of the customer organization, chart a course and have a crew to help reach the desired destination. Having a good map is critical.

Maintaining a skills map is an ongoing activity throughout the Run & Improve stage. It helps to maximize the skills and capabilities of people while enabling staff to undertake work that is aligned to their skills and aspirations (see **Figure 19: Communication skills map**). With such diverse teams in a SIAM ecosystem, it is necessary to understand the capabilities required to achieve optimal results.

Each service provider must identify the levels of capability and capacity they need to deliver its services and then consider the skills it has against what is needed. This will identify gaps that need to be addressed to avoid risks associated with not having the right level of capability available at the right time. Gaps can be because of insufficient depth of knowledge, insufficient capacity to cover the volume or hours of work, or because of single points of failure. This must be regularly reviewed, maintained and gaps acted on.

5.7.1.3 Ongoing training needs analysis and training plans

Competency frameworks represent the starting point for staff development and workforce planning initiatives in all layers. Continuing to develop staff can help organizations to stay competitive.

> **Training**
>
> Training can be described as the acquisition of skills, concepts or attitudes that result in improved performance within the job environment.

A training needs analysis (TNA) identifies training gaps by isolating the difference between current and future skills. A TNA looks at each aspect of an operational domain so that the initial skills, concepts and attitudes of the human elements of a system can be identified effectively and appropriate training can be specified. A TNA is the first stage in the training process and involves a procedure to determine whether training will indeed address the problem or gap that has been identified.

5.7.1.4 Succession planning

Succession planning is a strategic process that identifies critical roles, identifies and assesses possible successors, and provides them with the appropriate skills and experience for present and future opportunities. This facilitates the transfer of corporate skills and knowledge.

Succession planning provides a security net for the customer organization and protects it from risks that may result from service provider staff changes. To preserve organizational memory, it should be a deliberate and systematic effort designed to ensure continued effective performance of the SIAM ecosystem by making provision for the development and replacement of key people over time.

The service integrator should facilitate the transfer of skills and knowledge from service providers moving in and out of the ecosystem, to ensure sustained ways of working, and the right people with the right skills in the right place at the right time. The service integrator must identify those capabilities that are most critical to the success of the SIAM ecosystem, prioritizing succession risks and interventions accordingly. This approach needs to be based on the evolving needs of the customer organization, overcoming both structural rigidity and misalignment between strategic priorities and talent capabilities.

The hazards of stepping into service providers' responsibilities

In one case, a service provider was viewed as performing poorly on its change control process responsibilities. As a tactical step, the service integrator stood up a change management team and instructed the service provider to allow it to conduct change control.

When attempting to normalize the situation, the service integrator was told that the service provider team had redeployed the change managers, and there would be a cost to re-establish that team. Under the principle of estoppel, the service provider asked for funding to cover this cost.

There are three important phases of succession planning:

- **Mapping leadership roles and critical positions**: look beyond the basic skills and knowledge required to perform an adequate job and into the deeply rooted capabilities, such as traits and motives.
- **Define the parameters of critical positions**: the service integrator should create tools and templates to help identify critical roles. These should identify specific skills, capabilities, knowledge and qualifications required for success in all critical positions. This should lead to the development of a more comprehensive competency list based on staffing needs and associated risks.
- **Generate detailed position descriptions**: define the knowledge, skills and experience required for success for anyone assuming the role.

In addition, it is important to detail the type of learning and development that must be provided to train team members for these vacancies. This will serve as a learning curriculum, to support those moving into those roles.

Development opportunities within the curriculum may take the form of:

- Training
- Mentoring
- Shadowing
- Coaching

5.7.1.5 Improving processes

Most management frameworks emphasize measuring processes as a key element in ensuring the quality of their output and improving them. The effectiveness of the process is measured by comparing the output to the purpose. The minimum viable process should have everything defined to allow the most important measurements. Process lead time can be used as an efficiency metric.

A minimum viable process must have defined:

- Purpose
- Customers
- Triggers
- Outputs

During the Run & Improve stage, the service integrator needs to apply close (operational) governance to ensure that all service providers (including the service integrator itself) are complying with the process requirements and agreements, especially with respect to process integration (see section **5.4 Audit and compliance**).

Measurement is essential during Run & Improve. It should focus on value and effectiveness, as well as clear communication, demonstrated through cohesive team working, clear roles and responsibilities, effective communication and a positive working environment. Processes used in the right environment and for the right reason will streamline work and provide consistency.

During the Plan & Build stage, the service integrator will define process inputs and outputs. The service providers should continue to check the value of how they are performing processes by reviewing the outputs and evaluating steps for relevance and value, thus ensuring process step relevance. The approach should be to provide leaner processes that deliver required outputs.

Problem management example

The customer organization wishes to have problems managed to reduce the impact of incidents caused by them. The process model defines what the service provider must deliver as output – either an improved workaround or a definitive solution.

It defines to which party that process output must be delivered, often the people resolving incidents or the people making the changes to remove the causes of problems, and what triggers problem management – the criteria for deciding which problems to track or to investigate.

However, there is no fixed set of steps that are guaranteed to lead to a successful result. To paraphrase Tolstoy, each problem is unhappy in its own way. Although there are many possible methods useful to manage problems, it is up to the service providers and the service integrator to apply those methods in a flexible way, and appropriate to the desired results and the available resources.

All providers should be encouraged to simplify their approaches. The fewer steps and interactions there are, the easier it is to provide cohesion across service providers. A minimum viable process is a process that can achieve its purpose with the least possible amount of definition and elaboration.

Traditionally, there are many different process elements:

- Purpose of the process
- Process owner
- Activities to be performed, and in which order
- Trigger

- Inputs and outputs of the various activities (and of the process as a whole)
- Providers of the various inputs and the consumers of the various outputs
- Rules, policies and other constraints that should be respected in performing the process activities
- Resources required to perform the activities of the process
- Tools used by those resources to support the execution of the process
- Process roles and their responsibilities
- Mapping of the organizational structure to the process roles
- Process documentation
- Process metrics
- Expected levels of performance

Processes need to evolve over time to accommodate change and to check that non-value adding activities have not been introduced. Poor processes do more harm than good and lead to:

- Negative impact to business processes and outcomes
- Customer complaints regarding service
- User, customer and support staff frustration
- Duplicated or missed work
- Cost increases
- Wasted resources
- Bottlenecks

Reviews of process relevance and value are necessary, either as part of an ongoing improvement initiative or when issues arise.

The following steps offer an action plan for such a review:

1. Map the process
2. Analyze the process
3. Redesign the process
4. Acquire resources, if necessary
5. Implement and communicate change
6. Review the process

Map the process

Process models are designed during the Plan & Build stage (see section **3.1.4 Process models**). They should include a flowchart or a swimlane diagram for each sub-process, and show the steps in the process visually. Swimlane diagrams are slightly more complex than flowcharts but are better for processes that involve several people or groups. It is important to explore each process step in detail, as some processes may contain sub-steps that are unknown or assumed.

Analyze the process

Use your flowchart or swimlane diagram to investigate the issues within the process. Consider the following questions:

- Where do team members or customers get frustrated?
- Which of these steps creates a bottleneck?
- Where do costs go up and/or quality go down?
- Which of these steps requires the most time, or causes the most delays?

Techniques to trace a problem to its origin can be useful, such as value-stream mapping, root cause analysis, cause and effect analysis or the 'Five Whys'. Speak to the people who are affected by the process. What do they think is wrong with it? Which suggestions do they have for improving it? Try a workshop setting with appropriate stakeholders from all layers within the SIAM ecosystem, and continually consider the relevance and value of all processes.

Lean systems thinking

Lean thinking is a business methodology that aims to provide a new way of thinking about how to organize human activities to deliver more benefits to society and value to individuals while eliminating waste.

Lean thinking assesses the waste inadvertently generated by the way the process is organized, by focusing on the concepts of:

- Value
- Value streams
- Flow
- Pull
- Perfection

The aim of Lean thinking is to create a Lean enterprise, one that sustains growth by aligning customer satisfaction with employee satisfaction, and offers innovative products or services profitably while minimizing unnecessary over-costs to customers, suppliers and the environment.

Lean thinking seeks dynamic gains rather than static efficiencies. It is a form of operational excellence aimed at taking costs out of processes. This is relevant in a SIAM ecosystem where double handling and complications can make their way into processes, simply because of the complexity of interactions created from having multiple stakeholders.

Redesign the process

This activity involves re-engineering process activities based on the identified shortcomings. It is best to work with those who are directly involved in the process. Their ideas may reveal new approaches, and they are more likely to buy into changes if they have been involved at an early stage.

Make sure that everyone understands what the process is meant to do. Then, explore how problems identified in previous steps can be addressed. Note down everyone's ideas for change, regardless of the costs involved.

As a next step, narrow the list of possible solutions by considering how the team's ideas would translate to a real-life context. Conduct an impact analysis to understand the full effects of the ideas generated. Then, carry out a risk analysis and a failure mode and effects analysis to spot possible risks and points of failure within your redesigned process. Depending on the focus, there may be an opportunity to consider customer experience mapping at this stage.

These tests will help to demonstrate the full consequences of each proposed idea, so the end result is the right decision for everyone. Once the team agree on a process, create new diagrams to document each step.

It is a good idea to use a process forum to undertake this activity, or a working group if quick results are needed for an issue that has recently occurred and needs prompt action.

Acquire resources

Some resource and cost allocations will be within the scope of the service provider management team or the service integrator. If not, the resources and budget need to be agreed and acquired. This may require the production of an outline business case listing the arguments for how this new or amended process will benefit the SIAM ecosystem, as well as timescales, costs and risks.

Implement and communicate change

Usually, new ways of working will involve changing existing systems, teams or processes. Once approved, the change can commence.

Rolling out a new process could be managed as a project, especially if it affects multiple layers. Plans will support careful management (see section **3.2 Organizational change management approach**). Planning includes ensuring training is done at the appropriate level. For a minor change, a briefing note or even a discussion might be all that is required. If the changes are significant, formal training programs may be necessary.

Whoever is leading the implementation should allocate time for dealing with early issues and consider running a pilot first, to check for potential problems. It is also important to ensure that, in the initial weeks of operation, the new process is adopted, and staff do not revert to old ways of working.

Review the process

Few things work perfectly right from the start. After making any change it is good practice to monitor how things are going in the weeks and months that follow, to ensure that the change is performing in line with expectations. This monitoring will also allow the identification of issues as they occur. Make it a priority to ask the people involved with the new process how it is working, and what – if any – feedback they have.

5.7.1.6 Service integrator activities

The activities carried out by the service integrator in the Run & Improve stage will depend on the SIAM model in use within the ecosystem.

Typical activities include:

Table 9: Service Integrator Activities

Activity	Example
Major incident coordination	• Coordinating the investigations by multiple service providers • Communicating the status to users and stakeholders • Obtaining root cause analysis reports
Release planning	• Maintaining and publishing an integrated release plan with all providers' releases (where relevant to the SIAM model) • Identifying and planning for any potential clashes • Assuring integration testing of end-to-end services
Capacity planning	• Consolidating business demand forecasts • Maintaining and sharing an integrated capacity plan for the end-to-end services with service providers • Checking service providers' capacity plans to ensure timely provision of capacity
End-to-end monitoring	• Monitoring end-to-end services • Alerting service providers • Supporting investigation of major incidents and problems
Incident coordination	• Coordinating the investigations by multiple service providers • Communicating the status to users and stakeholders
Problem coordination	• Coordinating the investigations by multiple service providers • Communicating the status to users and stakeholders • Reviewing the priority with the business
Change management	• Managing the approval of high risk and high impact changes, and changes that affect multiple providers

5.7.2 Measurement practices

Within the Run & Improve stage, the focus is on end-to-end service delivery. End-to-end service measurement refers to the ability to monitor an actual service, not just its individual technical components or providers. Effective measurement practices support the performance management and reporting framework defined in the Plan & Build stage (see section **5.3 Ongoing performance management and improvement**).

Peter Drucker is often quoted as saying, *"you can't manage what you can't measure"*. In a SIAM ecosystem, objective measurements are very necessary to be able to hold all parties accountable.

Once in place, the SIAM ecosystem needs to be measured in terms of the outcomes delivered to the customer organization. Additionally, a framework for the assessment of the issues contributing to substandard performance must be used.

Metrics support decisions on how to improve the ability to meet the business goals. Metrics act as a guide showing the current state of a service, process or component. In this sense, KPIs are acting as decision-making indicators of performance, against a process or technology aspect serving the organization.

One of the challenges when setting objectives against metrics is the tendency to drift from 'managing by metrics' to 'managing metrics'. Focus on the importance of measuring collaborative outcomes – the 'sum of the parts'. Metrics are the measurements used for assessing the outcomes. When parties try to 'game the system' by achieving metrics without managing the underlying outcomes, undesirable behavior follows.

Ultimately, metrics are needed to perform two activities:

1. Allowing the measurement of aggregated outcomes, showing the links from the end-to-end service down to individual components and from components back up to the end-to-end service
2. Managing the behavior of the service providers to encourage them to be more collaborative and focus on the aggregate goal

In the context of SIAM, the service should meet the requirements of the customer organization and its customers and stakeholders. In general, customers consume the 'top level' end-to-end services delivering business outcomes, and so that is what they care about. The service elements that are grouped to deliver these end-to-end services must be measured to assess the source of any issue, in addition to measuring the end-to-end service. Nothing is gained from the components all meeting their targets if the end-to-end service does not.

The service model used during implementation to map accountabilities is used in the Run & Improve stage to understand the contribution of components to the end-to-end service(s).

Tools and processes such as configuration management show the links and dependencies between components, services and service providers. This data can be used to aggregate information, events and statistics about the end-to-end context and possible impact of the components.

The following are examples of measurements and targets that would help measure the value of the SIAM ecosystem and end-to-end services:

- Reduce the failure of critical service outages by x% each quarter

- Improve the performance of every service delivered by x% every quarter (continual improvement)
- Reduce the cost of managing technology by x% each quarter (exclusive of people)
- Increase the use of self-service by x% in those areas where this is appropriate
- 'Right first time' as a metric is imposed across the value stream with the goal of no defects, bugs or incidents passed downstream
- Mean time to restore (MTTR) – reduce the time to notice, alert, log, investigate, diagnose, resolve, close and confirm closure of incidents by x% per quarter for all priority one incidents
- MTTR – reduce the time to notice, alert, log, investigate, diagnose, resolve, close and confirm closure of incidents by x% per year for all priority two incidents or lower
- All changes must be capable of being rolled back or forward fixed within agreed timescales for each change type
- All changes must be version controlled including documentation, software, infrastructure
- All services will pass a continuity test annually or, if critical, every quarter
- Configuration management system must be accurate to within x%, measured quarterly
- No change deemed critical can go live without service integrator approval (automated or manual)
- All services provided (people, processes, architecture, software) will meet SIAM or corporate governance policies unless otherwise agreed
- Any reporting must be consistent and coordinated across the value stream

---

Everything is a priority one

One company had set a contract clause that defined that all incidents were treated as a priority one.

The implications of this were understood but there was a desire to drive the culture of *'right first time'* and *'never let a defect go live'*.

It took two years, but incident volumes reduced by more than 60 percent overall, customer satisfaction was no longer measured as it was so high and the cost of service last recorded had decreased by 18 percent.

The objectives for SIAM need to be synchronized with the overall customer organization objectives, focusing on both the short- and long-term business objectives of the commissioning organization. With this information, the service integrator can create targets that are aligned with the big picture. When targets are clear, it is easier to agree on what to measure. Metrics are an essential management tool. The right metrics provide the information to make qualified decisions based on facts.

Reports should be designed with care and consider what is important to the stakeholders at the time, but remember these requirements will change and reports may need to change as well.

Visual management

Visual management is the set of practices that allow individuals to see what is happening quickly, understand if there are any issues, highlight opportunities and act as a guide for improvement. Visual management is the ability of a system to quickly show the current status to anyone who stands and observes, using key indicators.

Visual management should use a tool that can display real-time information, such as production status, delivery status, process or technology status. It uses simple representations that everyone (including the customer organization, service integrator, service providers and other stakeholders) will find meaningful.

This approach requires information to be available to those doing the work in a timely fashion, displayed so everyone in the area understands it. The information and metrics provided are intended to drive decisions and actions, and there needs to be a clearly defined process for acting and obtaining support from relevant parties when needed. This aligns with the role of the process forums and working groups.

Visual controls cover more broadly how an ecosystem is operating. Examples could include the status of incidents and problems, a baseline of configuration information, and the flow of operational activity and performance. All service delivery elements can be displayed using such visual controls. This approach to measurement throughout the SIAM ecosystem results in a minimum viable product (MVP) of measurement. Despite the collaborative end-to-end focus for outcomes, measurement should not be incapable of revealing individual failure or success. Drill-down reporting should be available to maintain focus on end-to-end outcomes (see **figure 27** below).

Service	Trend	Last Mth	Target
Service A		96.4%	90%
Service B		73.2%	90%
Service C		98.1%	95%
Service D		100.0%	80%

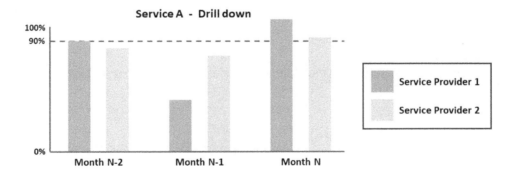

Figure 27: Drill-down reporting

The service integrator should maintain documentation that includes the measurement design and calculations to ensure clarity and transparency between all layers. This should include a statement of the intent of the measurement, which in turn enables a wider understanding and helps to provide a foundation for future improvement.

Outcome-focused measurements should be balanced with behavioral measurements, as measuring outcomes alone can drive unintended behavior. The relationship of related measurements should be defined, and where there are competing factors, use hierarchies or categories to articulate criticality. Balance the use of leading and lagging indicators, as each addresses the measurement of an outcome from a different perspective. Define the audiences for each measurement to ensure appropriate relevance and representation.

It is important to continue the journey of defining and documenting measurements throughout the Run & Improve roadmap stage, as knowledge is built and new measurements are identified to support continual improvement.

Visual management

A small organization with only limited experience was considering requirements to create a visual management solution. The organization wanted to keep things simple and gradually build on data.

They agreed that they wanted to:

- Build a simple view of the SIAM Run & Improve roadmap stage
- Highlight the service providers involved
- Visually highlight any issues
- Be capable of drilling down for more information
- Be adjustable to serve the needs of various viewers
- Be easy for all relevant service providers to participate
- Be flexible
- Only monitor information that showed the state of something
- Make the service integrator accountable for maintaining its view of the visual diagram
- Ensure that it was checked no less than every major release of an application, addition of a new feature or technical change

They defined some simple rules and created a value chain of activities. By bringing various factors together, they iteratively created a model that was flexible to use, change and apply against technology tool(s).

Hint – begin with something as simple as a series of large notes stuck to a wall and see if the flow works.

Figure 28 shows an example flow:

- Start high level
- Test
- Add detail
- Repeat

This can then be mapped to organization models, tools, roles, RACI matrices, etc. When mapping a process, ensure that the level of detail is consistent throughout the map. It is common to find some areas explored in detail and others at a high level, which can create ambiguity and inhibit action.

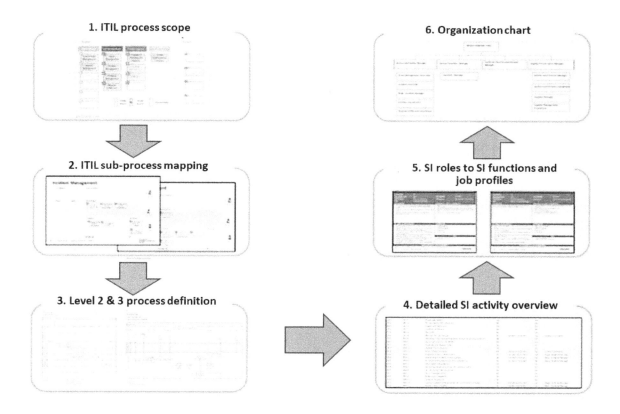

Figure 28: Visual management – ITIL process to organizational chart

Visual management provides a way of mapping the way the processes work to the organization, tools, metrics, roles, etc.

5.7.3 Technology practices

Within the Run & Improve stage of the SIAM roadmap, all layers need to keep abreast of emerging technology. The pace of change in the technology sector is accelerating rapidly, and organizations need to understand the potential of new technology.

There are several ways to keep abreast of new and upcoming technologies:

- Customer organizations often give early indication of a changing need or requirement. Much recent innovation is based on customer organizations expecting the same level of technology accessibility and functionality that they can get from their personal devices. It is not appropriate to ignore consumer technologies as not 'enterprise grade'. Customers often follow technology closely and are happy to provide their perspective on what is happening.
- It is the role of both the service providers and the service integrator to recognize new technology. For service providers that sell products, keeping up with the latest technologies helps maintain and improve their products, which better serves customers and meets changing needs, and enables market penetration and success. It is a good idea to encourage service providers to share their own product and service roadmaps as an input to future SIAM strategy. Recognize that this can have its challenges where information is commercially sensitive and relates to the service provider's market position and long-term business success.

- For service integrators, especially those that are external, keeping up with the latest trends and advising on them becomes a part of their added value.

Staying relevant in the technology industry is an ongoing challenge. There are some activities that will support the service providers and service integrator in staying abreast of emerging technologies. A technology assessment is the study and evaluation of new technologies. This is an important input into future business strategies and as such is a task for the customer organization to undertake or be involved with (see section **3.1.9 Tooling strategy**).

APPENDIX A: GLOSSARY OF TERMS

This glossary defines the terms used in this document. It expands the **SIAM Foundation BoK glossary** with additional terms.

Aggregation	Also referred to as **service aggregation**. Bringing together components and elements to create a group (or service).
Agile	A set of methods and practices designed to be applied for software development.
Agile retrospective	A meeting held at the end of an iteration, during which the team reflects on what happened and identifies actions for improvement going forward.
Agile SIAM	An approach to implement the core basic controls and structures of a SIAM model as a 'minimum viable product', with incremental improvements to realize rapid and regular value.
Association for Project Management (APM)	The chartered body for the project profession, developing and promoting project and program management.
Backlog grooming	Also referred to as **Backlog refinement**. When the product owner and some, or all, of the team review the backlog to ensure this contains the appropriate items, that they are prioritized and that the items at the top of the backlog are ready for delivery.
Balanced scorecard	A management system (not only a measurement system) that enables organizations to clarify their vision and strategy and translate them into action.
Benchmark	A standard or point of reference against which things may be compared.
Benefits realization management	A collective set of processes and practices for identifying benefits, aligning them with formal strategy and ensuring benefits are realized
Benefits realization plan	A document outlining the activities necessary for achieving the planned benefits. It identifies a timeline and the tools and resources necessary to ensure the benefits are fully realized over time.
Best of breed	A best of breed system or provider is the best system or provider in its referenced niche or category.

Better the devil you know	This is the shortened form of the full idiom 'better the devil you know than the devil you don't', and means that it is often better to deal with someone or something you are familiar with and know, even if they are not ideal, than take a risk with an unknown person or thing.
Blue/red/amber/ green reporting (BRAG)	See **red/amber/green reporting (RAG).** The addition of the B is to show completed items in blue, thus becoming a BRAG chart or report.
Board	Boards perform governance in the SIAM ecosystem. They are formal decision-making bodies, and are accountable for the decisions that they take. Boards are a type of structural element.
Boolean	A data type with only two possible values: true or false.
Business as usual (BAU)	The normal state of something.
Business case	Outlines a proposed course of action, its potential costs and benefits. Supports decision making.
Business process improvement (BPI)	A strategic planning methodology aimed at identifying the operations or employee skills that could be improved to encourage smoother procedures, more efficient workflow and overall business growth.
Capability	*"The power or ability to do something."*[36]
Capability assessment	Tools that help to identify and assess capability for current or future roles, and to plan for development needs.
Capital expenditure (CAPEX)	The funds that a business uses to purchase major physical goods or services to expand the company's abilities to generate profits. See also **Operational expenditure.**
Cloud services	Services that are provided over the internet, including software as a service (SaaS), infrastructure as a service (IaaS) and platform as a service (PaaS). Often treated as commodity services.
COBIT®	COBIT (Control Objectives for Information and Related Technologies) is a framework for IT governance and management created by ISACA.
Code of conduct	A code of conduct (or 'rules of the club') document is not a contractual agreement. It provides high-level guidance for how the parties in the SIAM ecosystem will work together.

---

[36] *www.lexico.com/definition/capability.*

Collaboration agreement	A collaboration agreement helps to create a culture based on working together to deliver shared outcomes, without continual reference back to contracts.
Commodity service	A service that can easily be replaced. For example, internet hosting is often a commodity service.
Common data dictionary	A central repository or tool that provides detailed information about the business or organization's data, and defines the standard definitions of data elements, their meanings and allowable values.
Common law	Sometimes called case law, court-made law or judge-made law. Law formed and developed by the courts rather than by a parliament.
Conflict of interest (CoI) plan	An instrument through which an organization seeks to eliminate, reduce or otherwise manage conflicts of interest or commitment.
Contract	An agreement between two legal entities. SIAM contracts are often shorter in duration than traditional outsourcing contracts, and have targets to drive collaborative behavior and innovation.
Cross-functional team	*"A group of people with different functional expertise working towards a common goal. It may include people from finance, marketing, operations and human resources departments. Typically, it includes employees from all levels of an organization."*[37]
Current mode of operation (CMO)	How things stand today; 'as-is'. See also **Future mode of operation.**
Customer (organization)	The customer organization is the end client that is making the transition to SIAM as part of its operating model. It commissions the SIAM ecosystem.
Cynefin™ (Pronounced ku-nev-in)	A Welsh word that signifies the multiple factors in the environment and considers how experience influences stakeholders. The Cynefin framework allows decision makers to see things from new viewpoints, assimilate complex concepts and address real-world problems and opportunities.
Dashboard	Provides 'at a glance' views of data, metrics, etc. relevant to a particular objective or business process.
Data room	A place where all information regarding an organization or situation is stored – such as a library – where people can go to learn about the organization or situation, and check facts and data.

---

[37] Wikipedia: *https://en.wikipedia.org/wiki/Cross-functional_team*.

Disaggregation	Splitting a group into component parts.
Early life support (ELS)	Last phase of a transition before handover to operation, during which specific support is provided for the implemented change.
Ecosystem	An ecosystem is a network or interconnected system. The SIAM ecosystem includes three layers: customer organization (including retained capabilities), service integrator and service provider(s).
Enterprise architecture	A definition of the structure and operation of an organization. It maps the current state and can be used to support planning for desired future states.
Enterprise process framework (EPF)	Mechanism for harmonizing process disparity and managing the associated complexities within a SIAM ecosystem.
Enterprise service bus	A type of 'middleware' that provides services to link more complex architectures.
Entity relationship diagram (ERD)	A graphical representation of an information system that shows the relationship between people, objects, places, concepts or events within that system.
Escalation	Raising the focus, seriousness or superiority of a subject.
Estoppel	Principle that precludes a person from going back on his or her original word or alleging facts that are contrary to previous claims or actions.
Ethical wall	Screening mechanism that prevents a conflict of interest by ensuring people and or organizations do not share certain information regarding another person or organization.
Exit services schedule	A contract schedule that prepares for the eventuality of service exit, defining handover activities to be performed by the outgoing service provider.
External service provider	An external service provider is an organization that provides services and is not part of the customer organization. It is a separate legal entity.
Externally sourced service integrator	Type of SIAM structure: the customer appoints an external organization to take the role and provide the capabilities of the service integrator.
Framework	A system of rules, ideas or beliefs that is used to plan or decide something, for example, the **ITIL** framework.

Function	*"An organizational entity, typically characterized by a special area of knowledge or experience."*[38]
Future mode of operation (FMO)	How things will stand after a transition period; 'to-be'. See also **Current mode of operation.**
Gaming the system	Also 'gaming the rules'. Using the rules and procedures meant to protect a system in order, instead, to manipulate the system for a desired outcome.
Governance	Governance refers to the rules, policies, processes (and in some cases, legislation) by which businesses are operated, regulated and controlled. There may be many layers of governance within a business from enterprise, corporate and IT. In a SIAM ecosystem, governance refers to the definition and application of policies and standards. These define and ensure the required levels of authority, decision making and accountability.
Governance board	Within SIAM, boards are regarded as **structural elements** that perform a key role in providing governance. They do this through acting as decision-making bodies, which are convened regularly throughout the operating lifespan of the SIAM model.
Governance framework	Within a SIAM ecosystem, this allows the customer organization to exercise and maintain authority over the ecosystem. It includes corporate governance requirements, controls to be retained by the customer, governance structural elements, segregation of duties, and risk, performance, contract and dispute management approaches.
Governance library	See **SIAM library.**
Governance model	Designed based on the governance framework and roles and responsibilities. Includes scope, accountabilities, responsibilities, meeting formats and frequencies, inputs, outputs, hierarchy, terms of reference and related policies.
Greenfield (site or operation)	Something that lacks constraints imposed by prior work.
Hangout	Usually a virtual forum to allow an informal exchange between various parties.

---

[38] IT Process Wiki. *https://wiki.en.it-processmaps.com/index.php/ITIL_Functions.*

Heat map	A heat map is a graphical representation of data where the individual values contained in a matrix are represented as colors. Heat maps are often used to summarize findings ranging from areas that require attention ('hot') to those that are well established/stable/mature ('cold').
Hybrid service integrator	Type of SIAM structure: the customer collaborates with an external organization to take the role of service integrator and provide the service integrator capability.
Incumbent	Current holder of a position.
Infrastructure as a Service (IaaS)	A type of cloud service that allows customers to access virtualized computing resources.
Insourcing	Sourcing from within the organization.
Intelligent client function	See **Retained capabilities.**
Interdependency	The dependence of two or more people or things on each other.
Interim operating model	An operating model setting out how to operate during a transition stage, using temporary measures until 'full' transition is completed. See also **Future mode of operation.**
Interim service plan	See **Interim operating model.**
Internal service provider	An internal service provider is a team or department that is part of the customer organization. Its performance is typically managed using internal agreements and targets.
Internally sourced service integrator	Type of SIAM structure: the customer organization takes the role of service integrator, providing the service integration capability.
ISO/IEC 20000	A service management system (SMS) standard. It specifies requirements for the service provider to plan, establish, implement, operate, monitor, review, maintain and improve an SMS.
ITIL®	ITIL (Information Technology Infrastructure Library) is the most widely accepted approach to IT service management in the world, and is a registered trademark of AXELOS Limited.
Kaizen	Japanese word meaning 'change for better'. Activities that continuously improve all functions and involve all employees from the top down.

Kanban	'Signboard' or 'billboard' in Japanese. A method for managing the creation of products with an emphasis on continual delivery while not overburdening the development team.
Kanban board	A visual representation that shows the status of items in a workflow.
Keeping the lights on	This idiom refers to the day-to-day running of a business or its operations.
Kepner-Tregoe problem analysis	Systematic method to analyze a problem and understand the root cause of the issue instead of making assumptions and jumping to conclusions.
Key performance indicator (KPI)	A metric used to measure performance. KPIs are defined for services, processes and business objectives.
Layers (SIAM layers)	There are three layers in the SIAM ecosystem: customer organization (including retained capabilities), service integrator and service provider(s).
Lead supplier service integrator	Type of SIAM structure: the role of service integrator is taken by an external organization that is also an external service provider.
Leading and lagging indicators	Two types of measurements used when assessing performance in a business or organization. Leading indicators are typically input oriented, hard to measure and easy to influence. Lagging indicators are typically output oriented, easy to measure but hard to improve or influence.
Lean (systems) thinking	A business methodology that aims to provide a new way to think about how to organize human activities to deliver more benefits to society and value to individuals while eliminating waste.
Liquidated damages	Also referred to as liquidated and ascertained damages. Damages whose amount the parties designate during the formation of a contract for the injured party to collect as compensation upon a specific breach (for example, late performance).
Management methodology	A management methodology describes methods, rules and principles associated with a discipline.
Man-marking	An undesirable and wasteful type of micro-management, where the customer checks the work of the service integrator constantly.

Master services agreement (MSA)	A master services agreement is a contractual document between two (or more) parties that outlines the responsibilities of both (or all) parties. This will apply to all subsequent schedules or SoWs. It is commonly used where a supplier provides multiple services, as it enables the addition, change and removal of services without affecting the general responsibilities.
Mutually exclusive and collectively exhaustive (MECE)	MECE (pronounced 'me see') is a way of segmenting information into sub-elements that are **m**utually **e**xclusive and **c**ollectively **e**xhaustive. Elements should 'exclude' each other, for example, be distinct, and should 'exhaust' the relevant field, for example, contain everything that belongs to it.
Model (SIAM model)	A customer organization develops its SIAM model based on the practices, processes, functions, roles and structural elements described within the SIAM methodology. Its model will be based on the layers in the SIAM ecosystem.
Multi-sourcing	Sourcing of goods or services from more than one service provider.
Multi-sourcing integration (MSI)	May be used as a synonym for **SIAM.**
OBASHI	A framework and method for capturing, illustrating and modeling the relationships, dependencies and dataflows between IT assets and resources (**O**wnership, **B**usiness processes, **A**pplications, **S**ystems, **H**ardware and **I**nfrastructure), and is a licensed trademark of OBASHI Ltd.
Offboarding	The process for exiting/removing service providers from the SIAM ecosystem in a controlled manner.
Onboarding	The process for bringing new service providers into the SIAM ecosystem in a controlled manner.
Open Systems Interconnection (OSI)	A reference model for how applications communicate over a network.
Operational expenditure (OPEX)	Expenses that are incurred on maintenance, and the running of assets generated through **CAPEX.** **See also Capital expenditure.**
Operational level agreement (OLA)	Within the SIAM context, OLAs are created between parties (for example, the service integrator and a service provider) to break down end-to-end service targets into detail and individual responsibilities.

Operational level framework (OLF)	A collection of all the **OLA**s and associated **operating level measurement**s for a particular SIAM ecosystem.
Operations manual	Sometimes referred to as a run book. Provides a summary of the contract, service, deliverables and obligations in layman's terms, and to support an understanding of process activities and objectives for day-to-day use.
Operating level measurement (OLM)	Operating level measurements are associated with **OLA**s. They are produced by deconstructing service level commitments, deliverables and interactions that involve more than one party, providing appropriate measurements for each part of an end-to-end process.
Organizational change management (OCM)	The process used to manage changes to business processes, organizational structures and cultural changes within an organization.
Outcome	The way a thing turns out; a consequence or end result.
Output	*"Quantity of goods or services produced in a given time period, by a firm, industry, or country."*[39]
Outsourcing	Procuring goods or services from an external organization.
Performance management and reporting framework	The performance management and reporting framework for SIAM addresses measuring and reporting on a range of items, including: • Key performance indicators (KPIs) • Performance of processes and process models • Achievement of service level targets • System and service performance • Adherence to contractual and non-contractual responsibilities • Collaboration • Customer satisfaction
Platform as a Service (PaaS)	A type of cloud service that allows customers to use virtual platforms for their application development and management. This removes the need for them to build their own infrastructure.
Practice	*"The actual application or use of an idea, belief, or method, as opposed to theories relating to it."*[40]

---

[39] Wikipedia: *https://en.wikipedia.org/wiki/Output_(economics)*.

[40] *www.lexico.com/definition/practice*.

Prime vendor	A sourcing approach where the service provider sub-contracts to other service providers to deliver the service, and the customer only has a contractual relationship with the prime vendor.
Process	A documented, repeatable approach to carrying out a series of tasks or activities.
Process forum	Process forums are aligned to specific processes or practices. Their members work together on proactive development, innovations and improvements. Forums will convene regularly, for as long as the SIAM model is in place. Process forums are a type of structural element.
Process manager	Responsible for process execution.
Process model	Describe the purpose and outcomes for a process, as well as activities, inputs, outputs, interactions, controls, measures, and supporting policies and templates.
Process modeling	Analytical illustration or representation of an organization's activity or set of activities intended to support the achievement of specific goals or outcomes.
Process owner	A process owner is accountable for end-to-end process design and process performance.
Product backlog	A prioritized list of work for the development team that is derived from the roadmap and its requirements. The most important items are shown at the top of the product backlog so the team knows what to deliver first.
Program management	The process responsible for managing groups of projects to deliver a unified goal.
Project management	A process that provides a repeatable approach to deliver successful projects.
Quality gates	A checkpoint (gate) at a certain point in time that assesses whether the project is still viable and on track to deliver its benefits.
Red/amber/ green reporting (RAG)	Reporting using traffic light coloring. Items going to plan are highlighted in green, items at risk of missing planned dates are highlighted in amber and items that have missed planned dates are highlighted in red.
Request for information (RFI)	A business process used to compare service providers by collecting information about them and their capabilities.
Request for proposal (RFP)	A business process used to allow service providers to bid for a piece of work or project.

Responsible, accountable, consulted and informed (RACI)	RACI is an acronym that stands for responsible, accountable, consulted and informed. These are the four principal 'involvements' that can be assigned to an activity and a role. A RACI chart is a matrix of all the activities or decision-making authorities undertaken in an organization, set against all the people or roles.
Responsibility	A responsibility is *"something that it is your job or duty to deal with."*[41]
Results chain	A qualitative diagram of the relationship between initiatives (inputs) and outcomes, and the contributions (connections) between the two.
Retained capability/ capabilities	The customer organization will include some retained capabilities: the functions that are responsible for strategic, architectural, business engagement and corporate governance activities. Retained capabilities are sometimes referred to as the 'intelligent client function'.
Risk management	The forecasting and evaluation of uncertainties, together with the identification of procedures to avoid or minimize their impact.
Roadmap	The SIAM roadmap has four stages: Discovery & Strategy, Design & Build, Implement, Run & Improve.
Role	A role is *"the position or purpose that someone or something has in a situation, organization, society or relationship."*[42]
Run book	See **operations manual.**
Scrum	An Agile framework that emphasizes teamwork, accountability and iterative progress toward a well-defined goal.
Scrum master	The facilitator for a development team who makes sure that Agile processes are being followed, and supports the product owner and the team.
Separation of duties/concerns	An internal control used to prevent errors or fraud. Separation of duties defines what each role is authorized to do and when more than one person must be involved in a task. For example, a developer might not be permitted to test and approve his or her own code.
Service	A system that meets a need, for example, email is an 'IT service' that facilitates communication.
Service aggregation	See **Aggregation**.

---

[41] *https://dictionary.cambridge.org/dictionary/english/responsibility*.

[42] *https://dictionary.cambridge.org/dictionary/english/role*.

Service assets	The resources and capabilities of an organization used to create value in the form of goods and services.
Service boundaries	A definition of what parts make up a service (what is 'inside the boundary'), often used in technical architecture documents.
Service consumer	The organization directly using the service.
Service credits	Or service level credits. A mechanism by which amounts are deducted from the amounts to be paid under the contract to the supplier if actual supplier performance fails to meet the performance standards set in the service levels.
Service dashboard	See **Dashboard.**
Service definition	A document intended to fully define the service and processes necessary to provide and support the service.
Service element	Or component. Part of a service that can be allocated to a specific service provider.
Service grouping	A collection of services.
Service improvement plan (SIP)	A plan or roadmap for improving service levels, often used within the continual service improvement process.
Service integration (SI)	May be used as a synonym for **SIAM.**
Service integration and management (SIAM)	SIAM is a management methodology that can be applied in an environment that includes services sourced from a number of service providers. Sometimes referred to as SI&M.
Service integrator	A single, logical entity held accountable for the end-to-end delivery of services and the business value that the customer receives. The service integrator is accountable for end-to-end service governance, management, integration, assurance and coordination.
Service integrator layer	The service integrator layer of the SIAM ecosystem is where end-to-end service governance, management, integration, assurance and coordination is performed.
Service line	A grouping of all the products and services related to one particular division of a business or one particular end-to-end service. For example, Apple has a service line for telephones and another one for personal computers.

Service management	The management practices and capabilities that an organization uses to provide services to consumers.
Service management and integration (SMAI)	May be used as a synonym for **SIAM**.
Service management integration (SMI)	May be used as a synonym for **SIAM**.
Service manager	Responsible for service delivery for one or more services.
Service model	A way of modeling the hierarchy of services, including services that are directly consumed by the customer organization and underpinning services and dependencies.
Service orchestration	The term used to define the end-to-end view of service activities and establishing the standards for inputs and outputs to the end-to-end process. This includes defining control mechanisms while still allowing service providers to define the mechanisms of fulfilment and the freedom to pursue internal processes.
Service outcomes	A definition of what a service is meant to achieve or deliver.
Service owner	A role that is accountable for the end-to-end performance of a service.
Service provider	Within a SIAM ecosystem, there are multiple service providers. Each service provider is responsible for the delivery of one or more services, or service elements, to the customer. It is responsible for managing the products and technology used to deliver its contracted or agreed services, and operating its own processes. They can be internal or external to the customer organization. Historically referred to as towers, or may also be referred to as vendors or service providers.
Service provider category	Service providers can be categorized as strategic, tactical or commodity.
SFIA®	The Skills Framework for the Information Age – a framework of internationally recognized descriptions for responsibility characteristics and skills practiced at various levels of responsibility and experience by people working in IT-related roles. Available from the not-for-profit SFIA Foundation, *www.sfia-online.org*.
Shadow IT	Shadow IT describes IT services and systems commissioned by business departments without the knowledge of the IT department (sometimes referred to as 'stealth IT').

SIAM ecosystem	Represents all interconnected parties within the **SIAM model**, across the **layers**: the **customer retained capabilities**, the **service integrator** and all the **service provider**s (both internal and external).
SIAM environment	See **SIAM ecosystem**.
SIAM governance lead	This is a senior role sitting within the customer's retained capabilities, primarily responsible for providing assurance regarding the implementation and operation of the SIAM strategy and operating model. See also **SIAM operational lead** role.
SIAM library	Repository to store information such as meeting minutes, contractual data, templates and other artefacts.
SIAM model	See **Model**.
SIAM operational lead	Responsible for leading and managing the overall operation of the SIAM ecosystem, providing direction and leadership, as well as acting as the 'escalation' point for any management issues. The SIAM operational lead will normally sit within the service integrator layer. See also **SIAM governance lead** role.
SIAM scorecard	A high-level scorecard defining and evaluating individual SIAM related KPIs aligned to the stated benefits measurements. The scorecard can also focus on KPIs and metrics aligned to the transition to the SIAM model.
SIAM structures	The four structures describe how the service integrator is sourced: internally, externally, from a lead supplier or as a hybrid.
Skills map	A diagrammatic representation that sets out all the main employability skills in a way that shows their relationships to each other. The innermost oval has primary or fundamental skills, the middle oval contains intermediate skills and the outer circle very specific skills.
Social network	A network of social interactions and personal relationships within SIAM. Often a dedicated website or other application that enables users to communicate with each other by posting information, comments, messages, images, etc.
Software as a service (SaaS)	A cloud service where software is paid for monthly as a subscription rather than purchased as a one-time payment.
Sourcing	The procurement approach an organization adopts. For example, sourcing services internally or externally. Adopting SIAM will affect how an organization sources services and the types of contracts it puts in place with service providers.

Stakeholder	A person or group of people that have a particular interest or are impacted by something.
Stakeholder map	A map to clarify and categorize the various stakeholders: • What the stakeholder groups are • Which interests they represent • The amount of power they possess • Whether they represent inhibiting or supporting factors • Methods in which they should be dealt with
Staring contest	A confrontation between two parties in which neither side is prepared to back down.
Statement of requirements (SoR)	A proposal outlining a business problem or an opportunity seeking funding and approval, and stating the requirements to do so.
Statement of work (SoW)	A formal document that defines the entire scope of the work involved, and clarifies deliverables, costs and timeline.
Strategy	*"A plan of action designed to achieve a long-term or overall aim."*[43]
Structural element	Structural elements are teams that have members from different organizations and different SIAM layers. They include: boards, process forums and working groups.
Subject matter expert (SME)	Or domain expert. A person who is an authority in a particular area or topic.
Supplier	An organization from whom the customer receives goods or services. See also **Service provider**.
Swimlanes	A visual element used in process flow diagrams to distinguish job sharing and responsibilities for sub-processes within an overall process. Each role has a 'swimlane' in the diagram.
Theory of Constraints (ToC)	A suite of management concepts developed by Dr Eliyahu Goldratt for identifying the most important limiting factor (for example, a constraint) that stands in the way of achieving a goal, and then systematically improving that constraint until it is no longer the limiting factor.
Tooling strategy	Defines what tools will be used, who will own them and how they will support the flow of data and information between the SIAM layers.

---

[43] *www.lexico.com/definition/strategy*.

Tower	See **Service provider**.
Town hall meeting	An informal meeting where topics are discussed in a relaxed atmosphere.
Training needs analysis (TNA)	The process of identifying the skills/competency gaps by isolating the difference between current and future skills/competency.
Transformation	The act of transforming or the state of being transformed. It is the less defined SIAM program that makes all changes in all roadmap stages to realize the full benefits of the SIAM model.
Transition	The process of change from one form, state, style or place to another. A SIAM transition takes the ecosystem from the previous, non-SIAM state to the start of the Run & Improve stage. Transition is a defined project with a start and end point.
Visual management	Visual management is the set of practices that allows individuals to see quickly what is happening, understand if there are any issues, highlight opportunities and act as a guide for improvement. Managing visually is the ability of a system to quickly show the current status to anyone that stands and observes, within 30 seconds, using key indicators that let everyone know how things are going.
War room approach	Using a single room to engage team members and stakeholders, driving collaboration, discussion and planning.
Waterfall	The method of development and implementation that follows sequential stages and a fixed plan of work, for example, plan, design, build and deploy.
Watermelon effect (Watermelon reporting)	The watermelon effect occurs when a report is 'green on the outside, red on the inside'. The service provider(s) meet individual targets, but the end-to-end service is not meeting the customer's requirements.
Win-win	A win-win situation or outcome is one that benefits both or all parties, or that has two distinct benefits.
Working group	Working groups are convened to address specific issues or projects. They are typically formed on a reactive ad hoc or fixed-term basis. They can include staff from different organizations and different specialist areas. Working groups are a structural element.

List of acronyms

This list expands the acronyms used in this document.

ADAM	Application development and management
ADKAR	Awareness, desire, knowledge, ability and reinforcement
AG	Aktiengesellschaft (German: Stock Corporation)
APM	Association for Project Management
BAU	Business as usual
BoK	Body of knowledge
BPI	Business process improvement
BRAG	Blue, red, amber, green
CAPEX	Capital expenditure
CCN	Contract change notice
CMMI	Capability Maturity Model Integration
CMO	Current mode of operation
COBIT®	Control Objectives for Information and Related Technologies
CoI	Conflict of interest
CSF	Critical success factor
CSI	Continual service improvement
EA	Enterprise architecture
ELS	Early life support
EPF	Enterprise process framework

ERD	Entity relationship diagram
EUC	End-user computing
EXIN	Examination Institute for Information Science
FAQ	Frequently asked questions
FCA	Financial Conduct Authority
FCR	First contact resolution
FMO	Future mode of operation
GDPR	General Data Protection Regulation
GM	General Motors
HR	Human resources
IaaS	Infrastructure as a Service
IP	Intellectual property
ISACA	Information Systems Audit and Control Association
IT	Information technology
ITIL®	Information Technology Infrastructure Library
ITO	IT organization
itSMF	Information Technology Service Management Forum
KPI	Key performance indicator
Ltd	Limited
MECE	Mutually exclusive and collectively exhaustive

MoR	Management of Risk
MSA	Master services agreement
MSI	Multi-sourcing integration
MVP	Minimum viable product
OBASHI	Ownership, business processes, applications, systems, hardware and infrastructure
OCM	Organizational change management
OLA	Operational level agreement
OLF	Operational level framework
OLM	Operating level measurement
OPEX	Operational expenditure
OSI	Open Systems Interconnection
PaaS	Platform as a Service
PMBOK	Project Management Body of Knowledge
PMI	Project Management Institute
PMO	Project management office
PRINCE2	PRojects IN Controlled Environments
RACI	Responsible, accountable, consulted and informed
RAG	Red/amber/green reporting
RCA	Root cause analysis
RFI	Request for information

RFP	Request for proposal
SaaS	Software as a Service
SCMIS	Supplier and contract management information system
SCT	Standard contract term
SFIA®	Skills Framework for the Information Age
SI	Service integration
SIAM	Service integration and management
SIP	Service improvement plan
SLA	Service level agreement
SMAI	Service management and integration
SME	Subject matter expert
SMI	Service management integration
SMS	Short message service (telephony) Service management system (ISO)
SoR	Statement of requirements
SoW	Statement of work
SOX	Sarbanes-Oxley
TNA	Training needs analysis
ToC	Theory of Constraints
TOGAF	The Open Group Architecture Framework
TOM	Target operating model

TUPE	Transfer of Undertakings (Protection of Employment)
UDE	UnDesirable Effect
UK	United Kingdom
USA	United States of America
WTO	World Trade Organization

APPENDIX B: CASE STUDIES

We have included some scenarios and examples from the real world. These case studies have been provided by members of the global author team involved in the development of this publication. They are based on personal experiences working with customer organizations and helping to develop and implement SIAM models across the globe.

Case studies help to emphasize contextual analysis of a number of events or conditions and their relationships. They are an excellent tool for bringing methodologies to life. They can enhance learning, simplify complex concepts, improve analytical thinking and develop tolerance on different views on the same subject.

When developing SIAM models, one size certainly does not fit all. These case studies should act as a reference for readers facing similar challenges within their own environments.

Case study 1 – Commercial airline, Australia

Background

In 2015, a transformation project was launched to implement a SIAM model within a large airline, headquartered in Australia. The airline, with approximately 35,000 employees worldwide, registers on average 125,000 incidents and 125,000 requests annually. 350,000 configuration items are known, though only half are active, with the remainder in a retired state. On average, 1,200 problems are registered and resolved each year, and the request catalog offers more than 1,000 products and services to the airline's employees.

To support this large, distributed and complex environment, approximately 200 service providers play a role in service delivery. Up to a dozen of these are major or 'tier 1' service providers, some with thousands of employees spread out globally via 'follow the sun' support models.

The multi-service provider network results in many challenges such as:

- Multiple languages, geographical locations and time zones
- Vastly differing cultures and ways of working
- Custom service management tools, processes and procedures for many service providers, with little automation and integration
- Lack of understanding of the business because of offshore support models

Drivers and challenges

In addition to the service provider specific challenges mentioned above, the airline's IT department had its own challenges including:

- Multiple 'first points of contact' for airline employees.
- Lack of confidence from employees that they would receive consistent and timely service, resulting in IT having a poor reputation.
- Silos within IT, which over time produced silos within its service providers.

- A watermelon effect, where service provider service levels were reported as green, but peeling back the layers showed deep dissatisfaction with the services and products provided. Because of the multiple service management tools (that were not always visible to IT), there was little confidence in the integrity of data being reported.
- A disjointed approach to service provider governance, and no clear method of analyzing and driving out improvements collectively.

Success factors

Despite the process and technology challenges, the success factors of the SIAM transformation project were primarily people focused, including:

- Changing the reputation of IT among employees.
- Building confidence from the airline's employees that IT was in fact adding value to the organization.
- Empowering service providers through 'trust and track' governance.
- Maturing thinking so that outcomes at all levels were more IT strategy aligned.
- Re-establishing a culture of integrity, respect, collaboration and good will. That is, while the contract was important as far as billing mechanisms were concerned, more focus was required on driving business outcomes, quality and improvements.

Partnering with the airline for the SIAM transition was an Australian-based managed service provider that was given the task of implementing a service centre and service integration function (with underpinning SIAM framework).

It was once said by the President of Ford Motor Company, Mark Fields, that *"Culture eats strategy for breakfast, lunch and dinner."*[44] As such, the primary driver for the partnership was specifically related to integrating the incumbent service providers' culture, values and ways of working with the airline. Deploying a different methodology, such as SIAM in and of itself would not have been enough. The airline required assistance in rebuilding the very fabric of its culture, in addition to the SIAM expertise that the service provider could bring.

Moving to a SIAM ecosystem

The SIAM ecosystem implementation was divided up into a number of stages, designed to ensure that appropriate planning and architecture activities occurred first.

Across the following stages, the business change team provided vital assistance in organizational change management (OCM):

- Planning
- Build and test
- Implement and improve

---

[44] *www.torbenrick.eu/blog/culture/organisational-culture-eats-strategy-for-breakfast-lunch-and-dinner*.

Discovery & Strategy

The SIAM implementation was underpinned by two primary methodologies, PRINCE2 and Agile. PRINCE2 ensured that appropriate planning and management products were in place, including business approval/endorsement via a steering committee. Once this was established, the Agile methodology enabled the project to then produce its deliverables in multiple sprints, inside stages.

Plan & Build

The sprints were applied to all aspects of build and test, such as hiring, documentation writing, software development and organizational readiness testing.

Despite its apparent age, Kotter's theory of change[45] was highly relevant for the implementation. Great emphasis was placed on ensuring that every aspect of planning, building and testing was conducted collaboratively with key stakeholders of the airline and its service providers. Workshops, town hall meetings, one on ones, newsletters, t-shirts and banners, standups and even yoga classes were used to ensure that the implementation would be as smooth as possible.

Key artefacts were endorsed for implementation during this phase, such as:

- IT governance framework and underpinning roles, responsibilities and charters.
- SIAM policies, processes, standards and controls.
- A central service management tool for primary processes and reporting.
- Training, work instructions, test plans/cases.
- A SIAM hybrid operating model that proposed a team comprising several process owners, operational managers and a supporting development team. The airline retained many governance activities, and as such, complementary roles were created within the airline itself so that the service integration function had a joint approach to service provider governance.
- Importantly, a relationship management framework ensured that regular discussions occurred regarding the importance of collaboration.

Implement, Run & Improve

Spread across three stages, service providers were onboarded into the new SIAM ecosystem according to the defined plans and artefacts built during previous phases. Because of the size and complexity of the changes, a big bang implementation approach was not suitable.

Following the full implementation, a number of benefits were realized:

- Cost reductions associated with elimination of multiple toolsets, multiple service desks and 'manual swivel chairing' of information between service providers.
- Immediate and lasting transparency of data and information within the airline's service management tool. This completely eliminated concerns regarding data integrity.
- Centralization and handover of ownership of intellectual property (IP) back to the airline.

---

[45] *www.kotterinternational.com/8-steps-process-for-leading-change*.

- Dramatic increase in customer satisfaction scores with airline employees and executives, from a baseline index of 55 to an average of 80 (out of a possible 100).
- Improvement in incident resolution, problem resolution and request fulfilment times. Through toolset integrations and modeling, average fulfilment times reduced from eight-and-a-half days to six-and-a-half days. The incident backlog was reduced by more than 500 records because of collaborative process forums, and visibility of data to the business improved.
- Strong alignment to IT strategy with measurable success based on IT key performance indicators (KPIs).

A number of improvements were identified and deployed as part of continual service improvement. Despite the OCM focus and training, there were gaps identified in knowledge, as well as the general understanding of the project.

The failure was identified in the abilities of those in leadership positions to pass the information further 'down the chain':

- Commercial constructs were not updated in full, which resulted in a lack of alignment across service providers. The service integration function governed and influenced service providers according to end-to-end requirements, however, the lack of contract updates reduced the speed of value realization.
- There was a general lack of understanding of how SIAM differed to enterprise service management (ESM). Service providers and the business often made assumptions that many activities previously executed by the ESM team would be completed by the service integrator. Additional documentation, awareness and SIAM Foundation training has improved understanding.
- The implementation of a central service management tool was a critical success factor (CSF) for the airline to improve data visibility and credibility. However, as a result, service providers decommissioned their service management processes and in some cases, entire roles were removed. This created a temporary vacuum at an ESM level, and the service integrator had to step in and bridge the gaps created. Through further improvements, the service integrator has moved back to a governance role.

Conclusion

A project is rarely implemented without its share of successes and failures. Thomas Edison, a famous inventor, was once quoted as saying, *"I have not failed. I've just found 10,000 ways that won't work"*.[46] The journey towards value realisation at this Australian-based airline was in some cases instant, and in others, hard won over many years. An article in 2014 by the 'IT Skeptic' Rob England cautions that any cultural change requires time, and the 'human rate of change' should be respected and embedded into expectations of success; *" ... good leaders do change culture (and bad ones), but it takes years. It is not an immediate fix"*.

---

[46] *www.brainyquote.com/quotes/quotes/t/thomasaed132683.html*.

As a methodology and concept, SIAM has completely changed the IT service management environment, enabling the airline's IT department to finally have the conversations that it needs to with its service providers and vice versa. This could only be achieved through careful planning, build/test, implementation and improvement.

Most important, though, is the ability to represent the culture and values that are required of the business and its service providers. Intentionality is required to ensure that the SIAM ecosystem rises above the 'noise' and exemplifies qualities of integrity, respect, collaboration and the values of the organization it is supporting. Without this, even the most sophisticated and advanced SIAM implementation will be *"eaten for breakfast, lunch and dinner"*.

Case study 2 – Logistics company, United Kingdom

Background

The following case study outlines the implementation journey to a SIAM ecosystem of a large logistics organization. The organization had a complex infrastructure and a diverse and very demanding customer base. In addition, market conditions were challenging. There were low costs and new entrants, but also established players acquiring smaller companies, adding to the risk of falling behind the competition and losing market share.

Drivers

The organization was coming to the end of a monolithic outsourcing deal, where all major IT processes had been outsourced to a single provider that had been the incumbent for seven years. IT leadership felt that the provider lacked control and oversight of many of the IT processes.

In addition, service quality was not consistent and the business units that consumed the IT services were concerned that IT was not well equipped to meet their current and future demands. Specifically, there was a need for agility and speed to market that the current operating model could not deliver. Finally, the risk of service outages affecting end-user satisfaction demanded robust processes for the prevention and recovery of incidents.

Implementing a SIAM model seemed like a good option. The ability to engage best-of-breed service providers to run specific areas of the IT operating model was very attractive to them. In addition, there was a need to move accountability for day-to-day service management activities back in-house.

Challenges

This desire to bring some aspects of service management in-house was seen as a critical design principle of the future operating model. Many years of outsourcing had resulted in a lack of personalization and innovation in the day-to-day running of the IT organization.

However, there were several hugely complex challenges:

- The capability to operate any processes in-house did not exist. This would need to be created and responsibilities transitioned from the incumbent service provider.
- There were no corporate tools for the support and delivery of IT services. A new service management tool was needed to automate the new processes and introduce workflows that could orchestrate the activities of the SIAM ecosystem and act as a single system of record.

- There were issues relating to the culture of the organization, specifically relating to the alignment with the business, which would need to be addressed.
- The skills required to operate these new processes and manage a complex SIAM ecosystem did not exist.
- The underpinning governance model to manage a SIAM ecosystem did not exist, so fundamental building blocks for innovation, service performance, process improvement and tooling strategy were not in place.

Moving to a SIAM ecosystem/Discovery & Strategy

A decision was made to move to a SIAM model. The strategic plan focused around two programs of work. One would handle the exit of the incumbent service provider and the procurement of the new sourcing contracts. The other would handle the design, build and implementation of the SIAM operating model.

Plan & Build

Critically, a single cross-program timeline was established that described milestones for key events, such as:

- Onboarding of the new service providers
- Contractual exit of the incumbent
- Service transition to the new ecosystem
- Definition of the SIAM operating model, including:
  o Definition of processes
  o Selection and implementation of tools
  o Enhancing capability
  o Development of the supporting governance model
  o Creation of the new organization

The SIAM operating model development focused upon the areas in italics above.

Definition of processes

Process definition workshops were run to capture requirements for the future state processes. Expert facilitation was engaged to visualize the future state operating model and the required elements of key processes that were not limited to core IT service management, but incorporated supplier and contract management, as well as the project delivery lifecycle.

Processes were documented, and process owners identified to take accountability for them. This included taking ownership of seeking buy-in and commitment to the processes. A process architect ensured that the processes were consistent and able to interact with one another effectively.

Selection and implementation of tools

A tooling strategy was needed to understand the existing landscape and define requirements for the future state. One major challenge was that many of the tools used by the organization for IT service management and project and portfolio management were owned by the incumbent service provider. These would either need to be transferred to the organization to manage, or replacements procured.

The existing tooling landscape was complex and relied heavily upon tools provided by the incumbent service provider that were shared across many of their customers. Clearly, ownership could not easily be transferred, and new tools would be required. Plans for rolling out tool selection projects were established.

Enhancing capability

The organizational design started with an initial assessment of staff numbers, roles and skills. A future state organization model was created, and roles were defined and sized using complex analysis of expected work volumes to determine how many staff were required.

Development of the supporting governance model

Governance processes and forums were developed to manage the following areas:

- Innovation
- Service provider performance
- Service reporting and review
- Tooling strategy and change
- Process model and change
- Contract management

Process owners were heavily involved in their development, and advice was sought from inbound service providers on how best to govern these elements in a complex SIAM ecosystem.

Creation of the new organization

This was based on an analysis of the current organizational model and the definition of a future 'to-be' state.

Implement

The SIAM operating model implementation focused upon:

- Implementation of processes
- Implementation of tools
- Implementing new capabilities
- Launching governance and process forums

Implementation of processes

Following the process definition workshops, the newly created process flows and newly engineered integration elements were rolled out. Identified process owners took on the accountability for and ownership of the process activities, and ongoing improvements.

Implementation of tools

Since many of the existing tools were owned by the incumbent service provider, replacements had to be procured. A tooling integration 'hub' was developed that enabled service providers to interface with the organization using their own tools if necessary. As service transition came closer, many service providers opted to use the customer organization's tools rather than their own, realizing the benefits of a single source of truth and a single set of defined workflow processes.

Implementing new capabilities

After the assessment of current staffing capabilities undertaken in the Plan & Build stage where the future state model was defined, an aggressive program of hiring was undertaken to address skills gaps. HR were engaged to manage individuals whose roles would change significantly as part of the change in IT operating model.

As role holders were appointed, they were immediately engaged to participate in process definition and tool selection activities. This was critical in obtaining buy-in and commitment to the new operating model, and critical in engendering a sense of ownership in the success of the program overall.

Launching governance and process forums

Process owners were heavily involved in the development of these and tasked with their ongoing improvement. The incumbent service provider transitioned its services to the new service providers. A program of knowledge transfer and testing was undertaken to provide assurance in the ability of the new service providers and the in-house service integration function to run the organization's critical services.

New tools were implemented for IT service management, service reporting and the project delivery lifecycle. These tools were critical in establishing a common workflow between service providers that contributed towards the feeling of operating as a single team.

Run & Improve

The 'one-team' ethos was a critical factor in the success of the program, which brought together practitioners from multiple organizations to form virtual process-aligned teams that shared a common workflow between service providers, which contributed towards the feeling of operating as a single team. This aligned to the common goal, to deliver and support the organization's existing and new services.

Conclusion

The SIAM model continues to operate successfully. IT is better prepared to meet the demands of the organization in terms of speed of delivery and service stability. When incidents occur, the retained service management team coordinates the resolution collaboratively with the SIAM ecosystem partners.

From a project delivery perspective, a single demand and project management process has been established to centralize all new work to be assessed and resources assigned from across the SIAM ecosystem.

The SIAM program is recognized as having been a success, and many of the principles applied have been replicated in other SIAM implementations.

Case study 3 – Global bank, Ireland

Background

This case study relates to a global Irish bank that moved to a fully outsourced model with a retained service integration layer.

Drivers

The existing service model was the result of a first-generation style outsourcing engagement with one incumbent provider that had been in place for one year.[47] Its role was to manage the service desk, end user computing (EUC) and specific applications within the environment. The provider was underperforming in the role and the bank was looking for an alternative.

Challenges

A major issue experienced was that the current service desk incumbent was underperforming. Additionally, the contractual terms, service level agreements (SLAs) and KPIs they were working to did not align to the contract model being deployed and adapted within the incoming service providers. This needed to be a consideration for the Plan & Build stage of the roadmap.

The main challenge to the organization was that it did not have the necessary experience of SIAM and multi-source environments. As such, it had to rely on outside consultancy and contract resources to design its future operational state, assist in the strategic sourcing and control, and govern and manage its transformation program.

The operational processes were immature and had to be redesigned to allow for a multi-source environment and for the overarching governance to work effectively. This issue was further compounded by the amount of other changes happening in the organization, and subsequently there was a great deal of effort engaged in managing that change activity, largely around perception management and changes to ways of working.

As the organization had limited experience of managing service providers on this scale, there were major challenges not only in defining the processes but also establishing them and ensuring that the teams responsible within the retained service integrator were capable of managing, governing and controlling these elements of the SIAM ecosystem. To assist in achieving this goal, workshops and learning sessions were instigated to mentor and train members of the retained organization in working with global partner organizations and how to work in collaboration and achieve benefit-based relationships with these working partners.

The whole transformation into the new SIAM world took 14 months, comprising 12 months to complete the two phases of the transformation and two months of parallel running. The Plan & Build team was in place to mentor, guide and assist the retained SIAM organization, but was also on hand to address and redefine any elements that required adjustment or change within the model.

---

[47] First generation outsourcing relates to the outsourcing strategies in the late 1990s and early 2000s.

Moving to a SIAM ecosystem/Discovery & Strategy

Having considered the options, the organization decided to further outsource in order to transform the IT organization from delivery focused to an assurance function via a retained service integration layer.

Plan & Build

In the first phase of the move to a SIAM ecosystem, three additional service domains were identified for the provision of data center services and hardware support, security and network provision. Two additional service domains were identified to supply application development and management (ADAM) for the second phase of the transition.

The transformation of the current IT organization (ITO) was planned alongside the procurement process identifying the service providers, to ensure that the service integration layer was in place and operational before the transition to the chosen service providers.

Because of the existing first-generation outsourcing arrangement, the SIAM element of this transformation was treated as a greenfield site. During the Plan & Build stage, the SIAM process suite and target operating model (TOM) could be developed and deployed cleanly within the new retained organization.

As indicated in the Challenges section, the current service desk provider was underperforming. Therefore, an additional element of the Plan & Build stage was created to establish a 'get-well' plan to ensure that the performance of the incumbent achieved the desired level. A contract change notice (CCN) was created and agreed to bring its contractual terms into alignment with the other providers within the SIAM ecosystem. This exercise proved to be very useful for the overarching SIAM Implement stage, as this provider became a 'testing ground' for the process suite before deploying it to the other providers. In addition, the retained capabilities could complete the scenario testing activity for the main processes, such as incident and major incident management, change, release and problem management, amongst others.

The existing operational processes, although in place, did not accommodate the new model, which would require them to be used across a multi-service provider ecosystem. In addition, there were no processes or functions in place to manage multiple providers from a supplier management, reporting, governance or financial management perspective, and subsequently these had to be defined and created to align to the SIAM model.

During the Plan & Build stage, a whole process suite was defined, taking into consideration the functional elements of the new structure. To consider that in parallel to the SIAM model deployment, a procurement exercise was proceeding for the incoming service providers. The process suite had to accommodate the retained service integration layer that would be required to interact, govern, control and provide management oversight across the ecosystem.

The retained elements of the organization had to be defined and designed from an organizational perspective, and aligned to the operational and assurance roles that had been created as part of the SIAM TOM and organizational plan. The impact of this was that certain delivery roles came under the TUPE legislation and other roles became redundant. This meant that the involvement of HR and organizational business change was required to manage this important and sensitive element of the transition.

Also in the Plan & Build stage, the decision was made that the Bank would retain ownership and control of the main service management tool and knowledge management repository. The current incumbent would be responsible for the management of the tool and act as custodian for when the incoming partners joined the SIAM ecosystem. The aim was to maintain the tool and knowledge repository as the central point of information (one version of the truth!). Also, when defining the knowledge repository, a governance library was created to manage all the collateral from the various governance boards and forums, management reporting, service reporting and associated OLAs and contract collateral for the SIAM ecosystem.

The parallel procurement phase for incoming SIAM providers was initiated alongside the SIAM Plan & Build stage, to ensure that the retained organization and its new model could be created and established in readiness for contract go-live with the new service providers.

As part of the initial phase of transition and due diligence, once the chosen providers had been selected and contractual terms agreed, process forums were initiated to align the providers with the new process suite, ways of working and governance model. In addition, collaboration working groups were set up where the partner providers could get to know their counterparts within the customer organization, the retained capabilities and the other service providers.

Implement

During the Implement stage, it was decided that the contract go-live dates for the various incoming service providers would be staggered to allow the internal service integrator to assimilate each provider into the new model, process suite, reporting and governance. Workshops were held to ensure that there was a clear understanding of the SIAM model and to assist with collaboration across the ecosystem. As each service provider was onboarded they would then join the next session with the service integrator to assist the next incoming partner.

Other parallel streams that were developed alongside the main SIAM team were the definition and deployment of demand management, business relationship management, partner management and financial management, and the alignment of these to the overarching governance and SIAM ecosystem.

The second main phase of the SIAM transformation was to align the ADAM providers into the ecosystem. To ensure consistency within the new 'one-team' approach, the same process forums and collaborative sessions were employed.

Run & Improve

This SIAM ecosystem has now been in place for two years and has been refined to enhance demand management as the patterns of business activity become clearer. The ADAM element has been enhanced by onboarding another provider, but also to accommodate for the replacement of an underperforming provider.

Conclusion

The new model is working well and developing through ongoing improvement activities.

APPENDIX C: LEAN, DEVOPS AND AGILE SIAM

SIAM is most commonly applied to IT and digital environments, products and services. The IT world changes rapidly, both in terms of technology, and also in terms of ways of working. This has led to suggestions that SIAM also needs to evolve, with terms such as 'Agile SIAM', 'Digital SIAM' and 'Lean SIAM' being used in presentations and discussion forums.

The SIAM Foundation BoK provides some considerations for applying various concepts and principles from DevOps, Agile and Lean in a SIAM ecosystem. The fundamental principles of SIAM do not change, but how service providers work together, the level of automation, governance considerations and the type of collaboration that takes place may need to adapt to embrace these changes.

This appendix provides an overview of how principles within Lean, Agile and DevOps can be considered within the SIAM roadmap and incorporated into the SIAM ecosystem. If the customer organization, service integrator or service providers already have proven capabilities in these areas, that can add value across the whole ecosystem. The service integrator will need to consider when, where and how to apply these ways of working.

SIAM and Lean

Lean principles are often employed by organizations seeking reduced costs and increased speed and quality for their products and services. Lean seeks to optimize flow and create a supportive culture focused on stability, standardization and reduction of waste. If your organization is already using Lean principles, it can add value in a SIAM ecosystem because of its focus on end-to-end value chains, collaboration, cooperation and flow.

Discovery & Strategy: Lean considerations

In the Discovery & Strategy phase of a SIAM transition, a clear vision is essential. In Lean terms that means applying the practice of Nemawashi and Hoshin Kanri, where influencers and stakeholders are prepared for change, and where the strategic decisions are well considered and communicated to stakeholders.

> **Nemawashi**
>
> Nemawashi is a Japanese term (根回し) for an informal process of quietly laying the foundation for a proposed change or project, by talking to the people concerned, gathering feedback and generating support. It is considered an important element in any major change, before any formal steps are taken, and successful Nemawashi enables changes to be carried out with the consent of all stakeholders.
>
> Nemawashi literally translates as "going around the roots". Its original meaning was literal: digging around the roots of a tree, to prepare it for a transplant.[48]

---

[48] For more information, visit: *https://kanbanzone.com/resources/lean/toyota-production-system/nemawashi/*.

Hoshin Kanri

Hoshin Kanri, another Japanese term (方針管理), also called Policy Deployment, is a method for ensuring that the strategic goals of an organization drive progress and action at every level within that organization. This eliminates the waste that comes from inconsistent direction and poor communication.

Hoshin Kanri strives to get every stakeholder pulling in the same direction at the same time. It achieves this by aligning the goals of the organization (strategy) with the plans of middle management (tactics), and the work performed by all employees (operations).[49]

For both Lean and SIAM environments, strategic intent should be based on customer value. When undertaking the analysis activities during this stage it is useful to apply Lean techniques, such as value stream analysis and root cause analysis. These will support landscape mapping and help to ensure that any 'as is' design impediments in processes and services are understood.

Plan & Build: Lean considerations

In the Plan & Build stage, organizational change management (OCM) activities begin. Like SIAM, Lean principles emphasize the importance of engaging with those who will be working within the SIAM model. As much as possible, stakeholders need to be actively involved in decision making about the ecosystem.

When designing the SIAM model, a Lean approach would consider simplifying workflows by minimizing handovers and avoiding long cycle times in process activities. In a SIAM model where processes need to be adapted to cater for complex models involving multiple service providers, this can be challenging. Throughout the Plan & Build activities, a focus on Lean principles will help to avoid over complication of activities, duplication and missed tasks, and optimizing workflows.

Implement: Lean considerations

When transitioning to a SIAM model both big bang and phased approaches are possible. Lean principles suggest a focus on single-piece-flow and small batches of work to avoid creating dependencies based on a deadline. Lean organizations would be likely to support a phased approach to transition, which allows for re-work between phases, and knowledge building and improvements.

Run & Improve: Lean considerations

Improvement has a high priority in organizations following Lean principles. Within a SIAM model, organizations can also utilize a Lean approach to improvements. If techniques such as Kaizen are already in use in an element of the SIAM model, they can be extended across the ecosystem.

---

[49] For more information, visit: *www.leanproduction.com/hoshin-kanri.html*.

Kaizen

Kaizen is a concept referring to business activities that continuously improve all functions and involve all employees from the C-suite to the operational personnel. Kaizen (改善) is the Japanese word for 'improvement'. Kaizen also applies to processes that cross organizational boundaries and, as such, may be useful within a SIAM ecosystem.[50]

Kaizen refers to business activities that involve all stakeholders, or layers within a SIAM model. It seeks to continuously strive to improve ways of working, and as such, the service integrator could use Kaizen techniques when working with the service providers and the customer organization.

Like SIAM, Lean supports principles such as 'fix first, argue later' and discourages blame assignment, instead encouraging effort to be focused on root cause analysis, building quality and removing waste.

One Lean principle, Kaikaku, also prepares an organization for more large-scale changes.

Kaikaku

Kaikaku (改革) is the Japanese term for 'radical change'. In business, Kaikaku is concerned with making fundamental and radical changes to a production system, unlike Kaizen, which is focused on incremental changes. Both Kaizen and Kaikaku can be applied to activities other than business as usual (BAU).[51]

Kaikaku is most often initiated by senior management. In SIAM, it could be initiated by a strategic or tactical board decision. It is intended to deliver significant impacts. Like other Lean approaches, Kaikaku is about introducing new knowledge, strategies, approaches and improvements to operational delivery. This scale of improvement is likely to be prompted by external factors, such as strategic change, market conditions and technology change. In a SIAM model, this could be linked to changes in the ecosystem, such as the addition or removal of a service provider.

Lean principles would support a SIAM ecosystem with a focus on standardization and waste elimination, without compromising on quality, cost and lead times.

Andon Cord

The Andon Cord is a concept from the Toyota Production System (TPS). Originally, it consisted of a cord or button that workers could use to 'stop the line' and halt production, to warn managers of an issue or potential issue. When the cord is pulled, managers and team members 'swarm' to the issue and carry out an investigation, and where necessary, resolution.

The Andon Cord is effective in a high-trust environment where workers feel safe and know they will not be blamed or punished.

---

[50] *www.kaizen.com/what-is-kaizen.html*.

[51] *https://en.wikipedia.org/wiki/Kaikaku*.

In a SIAM ecosystem, the Andon Cord would be a virtual concept, rather than a physical stop on a production line. Some multi-provider environments suffer because service provider staff do not feel confident about highlighting an issue, so manual workarounds or extra tasks are carried out unnecessarily. This can affect the quality of service received by the customer organization. The service integrator needs to work hard to create a high-trust culture where all service providers can feel confident to 'pull the cord', and where service providers work together to resolve issues that may be affecting them despite being outside of the scope of their contract.

SIAM and DevOps

Organizations utilizing an effective DevOps approach aim for a high degree of collaboration and mutual respect, as well as an appreciation of both the Dev and Ops considerations, goals and ways of working. Some organizations build multi-skilled teams including Dev and Ops competencies, and some keep the two disciplines separate and focus on collaboration.

DevOps values include a commitment to automation. Some organizations suggest that if a task is carried out more than twice, it should be automated (linked to the DevOps principle, "Improving daily work is more important than doing daily work"). This can also be applied to SIAM ecosystems, looking for ways to improve collaboration and integration between service providers.

Care must be taken, however, to build systems that still allow for service providers to be added and removed, when necessary, to retain the SIAM benefits of loose coupling.

The SIAM model needs to balance the benefits that DevOps ways of working can deliver with the complexity of the sourcing environment.

Discovery & Strategy: DevOps considerations

The governance model to support the SIAM ecosystem is an important output from the Discovery & Strategy stage of the SIAM roadmap. If the model also includes the use of DevOps principles, this approach will need to specifically consider boundaries and accountabilities. SIAM models often focus on the SIAM layers and the service groups, whereas DevOps maps structure at a technology or platform level, with links between application teams and platform teams.

Application teams often organize themselves using principles derived from Scrum or other Agile frameworks. In these structures, a hierarchy of decision points is designed to match the structure and architectural dependencies of the application. The SIAM governance model applied to DevOps teams needs to be aligned to current ways of working and decision making to preserve the autonomy of the teams.

Typically, DevOps teams have autonomy over a large portion of the application technology stack and processes, architecture and software deployment pipeline. They will typically engage in incident and problem management activities, as well as development activities. The SIAM model may need to reflect this way of working rather than forcing a rigid divide between Dev and Ops. This needs to be balanced against any separation of duties required for governance reasons. If the service desk is sourced by one of the external service providers in the SIAM model, consider how it will share knowledge with DevOps teams, and how DevOps teams will have visibility of issues.

In DevOps, application service provider teams own their product or service from inception to delivery. During Discovery & Strategy, this level of autonomy and accountability must be understood, and as much as is practical, be built into the model.

Plan & Build: DevOps considerations

If the SIAM model needs to align with DevOps ways of working, the Plan & Build stage is very important in establishing the key principles and concepts. The process models, governance model, collaboration model and tooling strategy all need to reflect how the ecosystem will work. The DevOps values of culture, automation, Lean, measurement and sharing can be used as a guide to help blend DevOps into the SIAM model.

Service providers that join the ecosystem need to have their DevOps capabilities assessed, and DevOps principles need to be covered during the onboarding process.

> **Autonomy versus control**
>
> Product and service teams that are used to working with DevOps need to be carefully integrated into SIAM models. The increased number of service providers involved in the ecosystem may mean that the team needs to share more information externally, which might be a new requirement contrary to the existing culture. If some teams continue to behave with complete autonomy and disregard for other service providers in the SIAM model, there can be negative consequences.
>
> For example, a case brought negative media attention to a service provider of payment services when all the ATMs in Austria went down. The service provider had not been informed about a change made by a DevOps team that was part of the same SIAM ecosystem. That change caused the failure of the ATM network, and significant reputation damage to the customer organization.

Implement: DevOps considerations

A phased approach to a transition to SIAM is more aligned with DevOps principles, but this will also depend on the real-world conditions, requirements, contract end dates, etc. For existing products and services developed and supported by DevOps teams, knowledge transfer and handover activities are essential. Small teams may have a great deal of knowledge that can be lost in the handover to another service provider.

Run & Improve: DevOps considerations

DevOps principles focus on feedback, and learning and experimentation. If these are already embedded in all or part of the SIAM model, they can be used to help support improvement activities. If one service provider, the service integrator or the customer organization has strong capabilities in these areas, the service integrator can focus on propagating this approach across the ecosystem.

SIAM and Agile

Many organizations still apply traditional command and control structures, with management hierarchies where goals and decisions flow from top to bottom. These organizations typically use linear planning and control mechanisms. In comparison, an Agile organization consists of cross-functional teams in a people-centric environment. It aims to drive value through iterative and incremental product and service development, with frequent feedback from customers. A SIAM model needs to enable cross-functional teams working across the layers, functions, processes and services. It is important to consider the impact of this Agile approach on the SIAM strategy, structure, processes, people and technology.

Discovery & Strategy: Agile considerations

Within the Discovery & Strategy stage of the SIAM roadmap, stakeholders can learn from and use Agile principles and techniques to support their activities. A SIAM transformation often follows an iterative and incremental path, with some activities repeated based on new information.

A SIAM model needs to foster trust, eliminate micro-management and devolve responsibility within the boundaries of an agreed governance framework. If the customer organization or service integrator has agile capabilities, they can use these to define roles and structures for self-governing teams that will fulfil these objectives. Agile leadership focuses on enabling, not controlling, through a culture of experimentation and learning. This integrates well with the SIAM culture of collaboration and innovation.

Plan & Build: Agile considerations

Both SIAM and Agile environments put a great deal of emphasis on the importance of effective cross-functional teams. A cross-functional team, such as a working group or forum, may include personnel from all the SIAM layers.

Agile approaches can be used to establish effective team dynamics, flow and knowledge sharing using improvement iterations to reduce complexity and improve the feedback loop to the customer.

The right balance between self-managing teams and governance is an important consideration in any environment, but especially so when considering the balance between an agile approach and the governance framework with a SIAM model.

Implement: Agile considerations

Self-organizing teams in Agile and SIAM

Self-organization is a fundamental concept in Agile methods and approaches. A common misconception is that because of the resilience of self-organizing teams, there is little or no role for governance and control. However, it doesn't mean letting people do whatever they want to do. Self-organizing does not mean no governance or control, it is defined and evident but exercised in a more subtle and indirect way.

Self-organizing teams are based on empirical control. The control is exercised by selecting the right people, creating an open environment, an evaluation system of group performance and encouraging providers to become involved early, not forcing or controlling them.

An Agile team's job is to self-organize around the challenges, and within the boundaries and constraints put in place by the organization and management, which varies from organization to organization and environment to environment. Individuals self-organize around a problem that is presented to them, but operate within defined boundaries. It is important when building a SIAM governance framework to ensure that this opportunity is not restricted.

Run & Improve: Agile considerations

SIAM structural elements can be integrated with Agile approaches and methods. Agile practices, such as standups, retrospectives, visual techniques, reviews and sprint planning, are activities that can be used to help working groups and process forums to be effective. For example, retrospectives can be used to assess process performance and ensure there is 'just enough' process in the SIAM ecosystem.

Agile uses tools and techniques, such as the product backlog and backlog grooming, to document and prioritize activities. Backlog items are discussed by the team, focusing on how important something is, how long it will take to accomplish and how success will be measured. These techniques can also be applied to SIAM transformations to help facilitate conversations between stakeholders and develop realistic plans.

It is important to have frequent reviews and showcases at the end of each Implement increment. This allows stakeholders to inspect and adapt based on early feedback. In a SIAM model, both process forums and working groups can support these activities.

APPENDIX D: STAFF DISPLACEMENT LEGISLATION

Within this publication we do not provide legal guidance. The intent here is to highlight the likely issues and provide examples of the considerations that might be required when designing and building the SIAM model.

Legislation, where available, is particular to a geography or country. Its purpose is to safeguard and protect the rights of employees in the event of a transfer of an undertaking, business or part of an undertaking or business to another employer as a result of a legal transfer or merger.

European legislation

There is a high-level of harmonization within European countries, but since European legislation is a directive, each member state has its own obligations. Some categories of workers are automatically protected, most fall under the directive and some are excluded.

Within Europe, the Acquired Rights Directive (ARD) exists. This is the name given to Directive 77/187 of 14 February 1977 (as amended and consolidated in Directive 2001/23 of 12 March 2001). It is implemented into national law in a variety of ways, for example:

- Statutory Regulation Transfer Regulations, Transfer of Undertakings (TUPE, UK)
- Statutory Regulation (French Labour Code L1224-1)
- Section 613 Civil Code and Works Constitution Act (Germany)

The directive influences the interpretation of national implementation so that employee rights (for example, sanctions) depend on the national employment framework.

United Kingdom

TUPE is a significant and often complex piece of legislation adopted by the UK. The purpose of TUPE is to protect employees if the business in which they are employed changes hands or is outsourced. Its effect is to move employees and any liabilities associated with them from the old employer to the new employer by operation of law.

TUPE protection applies to employees of businesses in the UK even though the business could have its head office in another country, but the part of the business that is transferring ownership must be in the UK.

TUPE means that:

- The employees' jobs usually transfer over to the new company – exceptions could be if they are made redundant or, in some cases, where the business becomes insolvent
- Their employment terms and conditions transfer
- Continuity of employment is maintained

The size of the business does not matter. TUPE is in place to protect all employees that are affected by a transfer.

TUPE definition

TUPE protects employees' terms and conditions of employment when a business is transferred from one owner to another. Employees of the previous owner when the business changes hands, automatically become employees of the new employer on the same terms and conditions. It is as if their employment contracts had originally been made with the new employer. Their continuity of service and any other rights are all preserved. Both old and new employers are required to inform and consult employees affected directly or indirectly by the transfer.

The Regulations were first passed in 1981, and amended in 2006, with further amendments made in 2014.

TUPE example

The IT department of a curtain-making business was transferred from a factory in Tamworth, UK (owned by company A) to company B based in Israel. None of the affected employees wished to transfer to work in Israel and so they were made redundant.

The employees' representing union brought a claim in the Employment Tribunal against company A for breaches of its obligations under TUPE to inform and consult employees and collective redundancy consultation laws. For this claim to succeed, the union needed to show that TUPE can apply to a transfer of a business outside of the UK (and, indeed, the European Union). Company A argued that it was not bound by obligations under TUPE as it could only apply to transfers between businesses within the EU.

At the time, there had not previously been a decision by the UK courts on this, although many commentators and practitioners had operated on the assumption that there was certainly a strong possibility that TUPE had extra-territorial application.

The European Appeal Tribunal (EAT) acknowledged that TUPE applies to a business transfer in cases where the undertaking is situated in the UK immediately before the transfer, and a service provision change in cases where there is an organized grouping of employees situated in the UK immediately before the service provision change.

Rejecting company A's arguments, the EAT agreed with the union that, in theory, TUPE can apply to cross-border transfers, even outside of the EU, despite the inherent difficulties in enforcing any consequential tribunal awards. Although acknowledging the common law presumption that legislation does not normally have extra-territorial effect, the EAT held that the UK courts had jurisdiction because employees can only avail themselves of rights under TUPE if the business was originally based within the UK. The purpose of TUPE is to protect employees upon the change of employer, and this applies to cross-border transactions.

This international element to TUPE was reflected in the service provision changes in 2006 which, the EAT stated, was "clearly aimed at the modern outsourcing of service provision, whether inside or outside the EU". The EAT referred the case back to the Employment Tribunal to closely examine the evidence to determine whether TUPE applied to transfer the employees to the transferee company.

This decision avoids the unscrupulous practice whereby a company can set up an overseas subsidiary to avoid the impact of TUPE. The EAT did note that it may be that, upon transfer of a business to a different country, the transferred business does not retain its identity as an economic

---

entity, negating TUPE's application in such cases. However, this point does not apply in relation to service provision changes and is judged on a case-by-case basis where it does apply.

This case raised legal and practical considerations for entities involved in international business transfers or off-shoring practices that fall within TUPE's scope, particularly those relating to call centers and IT support where UK-based services are often transferred to service providers abroad, either inside or outside the EU.

No doubt, the Acquired Rights Directive 2001, from which TUPE 2006 derives, will be revised to incorporate a definitive position on the cross-border application of its principles, allowing an opportunity for resolution of the complexities of extra-territorial application.

The Netherlands

In the Netherlands, the transferor and transferee generally have to consult the works council early enough for it to be able to influence the decision on whether and how to proceed with the transaction.

This involves at least one meeting with the works council, providing follow-up information, and considering the points it makes. If the works council is against the transfer, the transfer must be delayed by at least a month. The works council can also apply to the Enterprise Chamber, which can rule against the transaction going ahead.

France

In France, the sanctions for breaching consultation rules are criminal as well as civil. The criminal sanction is usually a fine though, theoretically, it can be a year's prison sentence – and the transaction may be put on hold until consultation is completed.

Both employers' works councils must be informed and consulted on the proposed transfer, and this process must be completed before any decision is made or any binding document is signed. Consultation runs for at least 15 days (though often collective agreements specify longer), then each works council has up to one month to issue its response. Works councils can delay, but not veto, any employee transfer.

Health and safety committees may need to be consulted, and sometimes prior authorization of the labor inspector is needed. Employees unlawfully dismissed as part of a transfer can achieve reinstatement as well as financial awards, and both employers can be held liable.

Considerations across the globe

Although most countries have no transfer laws, some have closely shadowed Europe's ARD, and others have introduced something that has similarities but radical differences too.

Below are some summary examples of transfer rules outside the EU. Of course, laws change regularly, and this is not legal advice. It is merely an indication of the current lack of consistency, leading to complexity and a requirement to act with caution and obtain the correct advice.

Singapore

In Singapore, the protection given to junior employees by the Employment Act has recently been extended and applies to many managerial and executive level staff too. These people have automatic transfer rights, similar to the ARD, so it is essential to work out who is covered, and who will not transfer and might therefore need to be recruited instead – perhaps receiving a contractual severance payment too.

In some ways, transfer laws are tougher on employers than in many EU countries, as the commissioner of labor can delay or prohibit a transfer, or can set conditions. Unlawful dismissals can lead to reinstatement, as well as compensation.

South Africa

South African law is similar to the ARD and unlawful dismissals can lead to reinstatement. Both employers have joint and several liabilities for a year after the transfer, and for events before the transfer. Consultation requirements are relatively mild, unless the employers want to agree a departure from the standard transfer laws.

Mexico

Mexican law uses the concept of a voluntary 'employer substitution letter' signed by the employee, or a procedure before the country's labor board, to achieve employee transfer. This applies in asset sales, but not outsourcing.

The employers share liability for six months after the transfer.

United States of America

Currently, the guidance covering the USA is very different to that in Europe. Regarding employee rights, there is no statute that governs the employment relationship when a business transfers to new ownership. As most employees are employed 'at-will', a new employer is free to offer employment to the employees of the seller/transferor employer or alter the terms and conditions of employment at the employment site.

In the instance where an organization is taken over by another company, there is no obligation for a party acquiring a business (an asset sale) to retain any of the seller's employees. However, if the new employer reorganizes the workforce after the transfer that results in a covered plant closing or mass layoff, the new employer or 'take over party' must provide the employees with 60 days' advance notice.

An employer who acquires a workforce consisting of unionized employees is required to bargain with the union in good faith regarding the effect of the layoff on unionized employees and, in certain situations, may be required to honor the terms and conditions of employment articulated in an existing collective bargaining agreement.

FURTHER READING

IT Governance Publishing (ITGP) is the world's leading publisher for governance and compliance. Our industry-leading pocket guides, books, training resources, and toolkits are written by real-world practitioners and thought leaders. They are used globally by audiences of all levels, from students to C-suite executives.

Our high-quality publications cover all IT governance, risk, and compliance frameworks, and are available in a range of formats. This ensures our customers can access the information they need in the way they need it.

Other resources you may find useful include:

- *Service Integration and Management (SIAM™) Foundation Body of Knowledge (BoK), Second edition, www.itgovernancepublishing.co.uk/product/service-integration-and-management-siam-foundation-body-of-knowledge-bok-second-edition*
- *ITIL® Foundation Essentials ITIL 4 Edition – The ultimate revision guide, second edition, www.itgovernancepublishing.co.uk/product/itil-foundation-essentials-itil-4-edition-the-ultimate-revision-guide-second-edition*
- *ITIL® 4 Essentials – Your essential guide for the ITIL 4 Foundation exam and beyond, second edition, www.itgovernancepublishing.co.uk/product/itil-4-essentials-your-essential-guide-for-the-itil-4-foundation-exam-and-beyond-second-edition*

For more information on ITGP and branded publishing services, and to view our full list of publications, visit *www.itgovernancepublishing.co.uk.*

To receive regular updates from ITGP, including information on new publications in your area(s) of interest, sign up for our newsletter at *www.itgovernancepublishing.co.uk/topic/newsletter.*

Branded publishing

Through our branded publishing service, you can customize ITGP publications with your organization's branding.

Find out more at *www.itgovernancepublishing.co.uk/topic/branded-publishing-services.*

Related services

ITGP is part of GRC International Group, which offers a comprehensive range of complementary products and services to help organizations meet their objectives.

For a full range of resources visit *www.itgovernance.co.uk.*

Training services

The IT Governance training program is built on our extensive practical experience designing and implementing management systems based on ISO standards, best practice, and regulations.

Our courses help attendees develop practical skills and comply with contractual and regulatory requirements. They also support career development via recognized qualifications.

Learn more about our training courses and view the full course catalog at *www.itgovernance.co.uk/training*.

Professional services and consultancy

We are a leading global consultancy of IT governance, risk management, and compliance solutions. We advise organizations around the world on their most critical issues, and present cost-saving and risk-reducing solutions based on international best practice and frameworks.

We offer a wide range of delivery methods to suit all budgets, timescales, and preferred project approaches.

Find out how our consultancy services can help your organization at *www.itgovernance.co.uk/consulting*.

Industry news

Want to stay up to date with the latest developments and resources in the IT governance and compliance market? Subscribe to our Weekly Round Up newsletter and we will send you mobile-friendly emails with fresh news and features about your preferred areas of interest, as well as unmissable offers and free resources to help you successfully start your projects. *www.itgovernance.co.uk/weekly-round-up*.

EU for product safety is Stephen Evans, The Mill Enterprise Hub, Stagreenan, Drogheda, Co. Louth, A92 CD3D, Ireland. (servicecentre@itgovernance.eu)

www.ingramcontent.com/pod-product-compliance
Lightning Source LLC
Chambersburg PA
CBHW081227050326
40690CB00013B/2689